T0207456

# Ethics of Science and Technology Assessment

Volume 44

Schriftenreihe der EA European Academy
of Technology and Innovation Assessment GmbH
*edited by Petra Ahrweiler*

More information about this series at http://www.springer.com/series/4094

Margret Engelhard
Editor

# Synthetic Biology Analysed

## Tools for Discussion and Evaluation

 Springer

*Editor*
Margret Engelhard
EA European Academy of Technology
    and Innovation Assessment GmbH
Bad Neuenahr-Ahrweiler
Germany

*Series editor*
Prof. Dr. Petra Ahrweiler
EA European Academy of Technology
    and Innovation Assessment GmbH
Bad Neuenahr-Ahrweiler
Germany

ISSN 1860-4803                    ISSN 1860-4811    (electronic)
Ethics of Science and Technology Assessment
ISBN 978-3-319-79741-0        ISBN 978-3-319-25145-5    (eBook)
DOI 10.1007/978-3-319-25145-5

Printed on acid-free paper

This Springer imprint is published by SpringerNature
The registered company is Springer International Publishing AG Switzerland

# EA EUROPEAN ACADEMY
## OF TECHNOLOGY AND INNOVATION ASSESSMENT

## The EA European Academy

The EA European Academy of Technology and Innovation Assessment GmbH deals with the relation of knowledge and society: Science, technology and innovation change our societies rapidly. They open new courses of action and create opportunities but also introduce unknown risks and consequences. As an interdisciplinary research institute, the EA European Academy analyses and reflects these developments. The EA European Academy was established as a non-profit corporation in 1996 by the Federal German state of Rhineland-Palatinate and the German Aerospace Center (DLR).

## The Series

The series Ethics of Science and Technology Assessment (Wissenschaftsethik und Technikfolgenbeurteilung) serves to publish the results of the work of the European Academy. It is published by the academy's director. Besides the final results of the project groups the series includes volumes on general questions of ethics of science and technology assessment as well as other monographic studies.

## Acknowledgement

The study "Synthetic biology analysed. Tools and arguments for discussion" was supported by the Klaus Tschira Stiftung gGmbH.

# Preface

An interdisciplinary approach is not only a key characteristic of synthetic biology but has also been the endeavour of our work in analysing this new field. This book is the result of a joint effort of eight researchers from fields as different as synthetic biology, xenobiology, microbiology, philosophy, ethics, social science and law. The project group met fourteen times over a period of 3 years to discuss and frame the topic, and to introduce each other to the individual perspectives on synthetic biology. Getting to know these manifold viewpoints helped to develop an integrated perspective presented here. Complex as synthetic biology is, this book should be seen as an interdisciplinary toolbox to facilitate and differentiate its societal evaluation. Members and authors of this study are Michael Bölker (Marburg), Nediljko Budisa (Berlin), Kristin Hagen (Bad Neuenahr-Ahrweiler), Christian Illies (Bamberg), Georg Toepfer (Berlin), Gerd Winter (Bremen) and Margret Engelhard (Bad Neuenahr-Ahrweiler), who chaired the group. The group received, in addition, input from external speakers whose contributions enriched the study, and in this respect the authors' special thanks go to Roger Brownsword (London), Arnim von Gleich (Bremen) and Christoph Then (Munich).

I would like to take the opportunity to thank the members of the group for their openness and commitment to sharing their ideas and realizing this publication. The project has been generously funded by Klaus Tschira Stiftung gGmbH. I express my gratitude to Irene Rochlitz (Hertfordshire) for help with the editing and to Arne Willée from the European Academy for his reliable assistance with the project. I also thank the people who helped the group in organising the various meetings that took place outside the Academy Bamberg, Berlin, Bremen, Marburg and Madrid.

Last but not least I am especially grateful to my family, Alena and Ilja, for their constant support!

Bad Neuenahr-Ahrweiler                                                                     Margret Engelhard
August 2015

# Authors of This Study

The whole content of the present volume has been discussed and integrated collectively by all authors of this study. The results reflect the intense interdisciplinary discussion of the working group and grew out of this work. However, individual authors took primary responsibility for single chapters as follows:

Chapter 1: all authors
Chapter 2: Michael Bölker, Margret Engelhard, Nediljko Budisa
Chapter 3: Margret Engelhard, Michael Bölker, Nediljko Budisa
Chapter 4: Georg Toepfer
Chapter 5: Christian Illies
Chapter 6: Rafael Pardo, Kristin Hagen
Chapter 7: Gerd Winter

# Contents

# Editor and Contributors

## About the Editor

**Margret Engelhard** studied biology in Marburg and Edinburgh and graduated in 1997 in micro- and molecular biology from the Phillipps-Universität Marburg and the Max-Planck Institute for Terrestrial Microbiology. Thereafter, she continued to work in wet-science and was conferred a doctorate in biology from the University of Basel. For the genetic aspects of this work she stayed for half a year at the University of Geneva. The focus of her research during her diploma and Ph.D. were nitrogen fixation in plant–microbe interactions. From 2004 to April 2015 Dr. Engelhard was member of the scientific staff of the EA European Academy of Technology and Innovation Assessment GmbH and coordinated, initiated and led interdisciplinary research projects with a focus on technology assessment in medical ethics and biotechnology, namely on organ transplantation, pharming and synthetic biology. Since then she works at The German Federal Nature Conservation Agency (Bundesamt für Naturschutz) in the area of GMO regulation and biosafety.

## Contributors

**Michael Bölker** studied Biochemistry in Tübingen and Berlin and graduated in 1991 from the Freie Universität Berlin and the Institut für Genbiologische Forschung Berlin GmbH. From 1992 to 1996 he was Postdoc at the Institute for Genetics and Microbiology of the Ludwig-Maximilians-Universität München (habilitation in Genetics in 1996). Since 1997 he is professor for Genetics at the Faculty of Biology of the Philipps-Universität Marburg. He is member of the Marburg LOEWE Center for Synthetic Microbiology (SYNMIKRO). His research focuses on investigating the molecular and cellular biology of the plant pathogenic fungus *Ustilago maydis*, with an emphasis on the role of small GTPases during morphogenetic transitions. In addition, his lab studies the genetic and biochemical basis of secondary metabolism in *U. maydis*. Recently, he discovered cryptic peroxisomal targeting of metabolic enzymes via ribosomal readthrough of stop codons or differential splicing both in fungi and in animals.

Postal address: Philipps-Universität Marburg, Fachbereich Biologie, Karl-von-Frisch-Str. 8, 35032 Marburg, Germany

**Nediljko Budisa** received the Ph.D. degree in 1997 from Robert Huber. He was in the time period between 1993 and 2010 at the Max Planck Institute of Biochemistry in Martinsried (Germany) as a Ph.D. student, postdoc, assistant professor and independent research group leader. After vocation in 2008, he holds the Chair of Biocatalysis at TU Berlin from May 2010. In 2004, he received the BioFuture prize; he is the single-handed author of the book "Engineering the Genetic Code" and one of the pioneers of this area of research. His research, in the core of the Synthetic Biology, is aimed to provide a solid basis for laboratory evolution of synthetic life forms with novel chemical possibilities.

Postal address: TU Berlin, Department of Chemistry, Müller-Breslau-Straße 10, D-10623 Berlin, Germany

**Kristin Hagen** studied biology, philosophy and agricultural sciences at the University of Tromsø, Norway, and was subsequently a postgraduate student with Prof. Broom at the Department of Clinical Veterinary Science, University of Cambridge, where she obtained her Ph.D. in 2001 with a thesis entitled "Emotional reactions to learning in cattle". She went on to do postdoctoral research on cattle housing, behaviour and welfare at the University of Veterinary Sciences, Vienna, and at the Freie Universität Berlin. Since 2006, she has been involved with technology assessment at the EA European Academy of Technology and Innovation Assessment, focusing on ethics and governance of emerging biotechnologies: "pharming", genetic engineering in livestock, concepts of animal welfare, large animals as biomedical models and synthetic biology.

Postal address: EA European Academy of Technology and Innovation Assessment GmbH, Wilhelmstr. 56, 53474 Bad Neuenahr-Ahrweiler, Germany

**Christian Illies** holds the Chair for Philosophy (Ethics) at the University Bamberg since 2008. After his studies of biology and philosophy (*Diplom-Biologe* Konstanz 1989), he continued his studies as a Rhodes-Scholar in Oxford (Magdalen College; DPhil in 1995). From 1995 to 2002 he worked as an Assistant-Professor at Essen University (*Habilitation* at the Technical University Aachen 2002). At the same year, he became *Universitairdocent* of philosophy and ethics of technology at the Technical University Eindhoven where he co-organized the Center of Excellence for Ethics in Technology of the three Dutch technical universities. From 2006 to 2008, Christian Illies was KIVI-NIRIA Professor for Philosophy of Culture and Technology at the Technical University Delft. Visiting Scholarships/Professorships at University of Notre Dame/Indiana (1999), European College of Liberal Arts (ECLA) (2001–2002), University of Cambridge (2007, 2011/2012, 2015), Northern Institute of Technology/TU Hamburg-Harburg (annually since 2009), Nanjing University of Aeronautics and Astronautics/China (2014). His research focuses on ethics and meta-ethics, philosophy of biology, philosophical anthropology, and philosophy of culture and technology, in particular architecture.

Postal address: Chair of Philosophy 2, Otto Friedrich University, 96045 Bamberg, Germany

**Rafael Pardo-Avellaneda** is Professor of Sociology and has been the General Director of the BBVA Foundation since 2000. From 1996 to 2000 he was Professor of Research at the National Council for Scientific Research (CSIC), and Professor of Sociology at the Universidad Pública de Navarra from 1993 to 1996. He has held appointments at Stanford University as Visiting Professor (1998) and Visiting Scholar (1990, 1991, 1996), and as Visiting Scholar at the Massachusetts Institute of Technology (1987, 1988). He has been an advisor to General Directorate XII (Science and Technology) of the European Commission and to Intel Corporation in San José, USA. From 1994 to 1996, he chaired the National Evaluation Commission for Social Science Research Projects, part of the Spanish National Agency for Evaluation and Science Policy. His major research and publications deal with organization studies, innovation, scientific and technological culture, social dimensions of Artificial Intelligence, cloning, and environmental culture and values.

Postal address: Fundación BBVA, Paseo de Recoletos, 10, 28001 Madrid, Spain

**Georg Toepfer** is head of the department "Knowledge of Life" at the Centre for Literary and Cultural Research (ZfL) in Berlin. He studied biology and philosophy and received his diploma in biology from the University of Würzburg, his Ph.D. in philosophy from the University of Hamburg and his habilitation in philosophy from the University of Bamberg. Before starting at the ZfL he was part of the Collaborative Research Centre "Transformations of Antiquity" at the Humboldt-University Berlin. His principal area of research is the history and philosophy of biology, with a special focus on the history and theoretical role of basic biological concepts.

Postal address: Zentrum für Literatur- und Kulturforschung, Schützenstraße 18, D-10117 Berlin, Germany

**Gerd Winter Dr. iur. Dr. iur. h.c. (Luzern), Lic. rer. soc.**, is Professor of Public Law and the Sociology of Law at the University of Bremen. He obtained the first and second state examination in law after studies in Würzburg, Freiburg, Lausanne, and Göttingen, and a Licentiat in sociology at Konstanz. After spending a year as visiting scholar at Yale university, he was in 1973 appointed professor of law at the University of Bremen where he in 1994 founded the Research Center for European Environmental Law (FEU) of which he has since been director. He retired in 2008 but was entrusted a research professorship. He acted as a legal consultant in many court proceedings and political committees concerning environmental law. He also consulted on administrative and environmental law development in various developing and transition countries. His numerous publications are concerned with almost all branches of environmental law at national, European and international levels. They are accessible via http://www-user.uni-bremen.de/gwinter/veroeffchronol.html

Postal address: Universität Bremen, Fachbereich Rechtswissenschaft, 28353 Bremen, Germany

# Chapter 1
# The New Worlds of Synthetic Biology—Synopsis

## A diverse and dynamic field that should not be judged as a whole but rather by its specific new features.

Margret Engelhard, Michael Bölker, Nediljko Budisa, Kristin Hagen, Christian Illies, Rafael Pardo-Avellaneda, Georg Toepfer and Gerd Winter

**Abstract** Synthetic biology is a young and heterogeneous field that is constantly on the move. This makes societal evaluation of synthetic biology a challenging task and prone to misunderstandings. Confusions arise not only on the level of what part of synthetic biology the discussion is on, but also on the level of the underlying concepts in use: concepts, for example, of life or artificiality. Instead of directly reviewing the field as a whole, in the first step we therefore focus on characteristic features of synthetic biology that are relevant to the societal discussion. Some of these features apply only to parts of synthetic biology, whereas others might be relevant for synthetic biology as a whole. In the next step we evaluate these new features with respect to the different areas of synthetic biology: do we have the right words and categories to talk about these new features? In the third step we scrutinize traditional concepts like "life" and "artificiality" with regard to their discriminatory power. Lastly, we utilize this refined view for ethical evaluation, risk assessment, analysis of public perception and legal evaluation. This approach will help to differentiate the discussion on synthetic biology. By this we will come to terms with the societal impact of synthetic biology.

M. Engelhard (✉) · K. Hagen
EA European Academy of Technology and Innovation Assessment GmbH,
Wilhelmstraße 56, 53474 Bad Neuenahr-Ahrweiler, Germany
e-mail: engelhar@hs-mittweida.de

K. Hagen
e-mail: kristin.hagen@ea-aw.de

M. Bölker
Philipps-Universität Marburg, Fachbereich Biologie, Karl-Von-Frisch-Str. 8,
35032 Marburg, Germany
e-mail: boelker@uni-marburg.de

## 1.1 Attraction of Synthetic Biology

Synthetic biology was attractive to scientists, practitioners, policy makers and societal groups right from the beginning. This is mainly because synthetic biology aspires to rationally *design* the realm of the living according to our needs, rather than to rely on the products of natural evolution. It promises to fulfill many goals of "classic" genetic engineering but in a quicker, cheaper and more predictable way. Synthetic biology sets off to create[1] artificial cells and living machines, to reassign the genetic code or to change the basic chemistry of the genetic material. Such visions reflect both scientific curiosity and technological ambitions; they appeal to policy makers' searches for the next big step in converging technologies and associated economic prospects. Synthetic biology is fascinating; approaching the limits of life by controlling its code and synthesis seems to open up limitless possibilities. There is a considerable element of hype generated by some leading figures in the field, both as a way of building identity and of giving salience to the emerging area and, obviously, attracting support.

Synthetic biology has also become a topic of consideration for philosophy and for science and technology studies: what are its ethical implications and how should it be governed? Early ethical and societal research about synthetic biology was partly invited by the field's protagonists,[2] who initially proposed strategies for

---

[1]The term "create" is used by a number of protagonists of the field, as well as "make", "redesign", "construct". We discuss the normative level of these expressions in Chap. 5 of this book.

[2]E.g. Craig Venter's ethics group (Cho et al. 1999); SynBerc Human Practices (Rabinow and Bennett 2009).

---

N. Budisa
Department of Chemistry, TU Berlin, Müller-Breslau-Straße 10, 10623 Berlin, Germany
e-mail: nediljko.budisa@tu-berlin.de

C. Illies
Chair of Philosophy 2, Otto Friedrich University, 96045 Bamberg, Germany
e-mail: christian.illies@uni-bamberg.de

R. Pardo-Avellaneda
Fundación BBVA, Paseo de Recoletos, 10, 28001 Madrid, Spain
e-mail: rpardoa@fbbva.es

G. Toepfer
Zentrum Für Literatur- Und Kulturforschung, Schützenstraße 18, 10117 Berlin, Germany
e-mail: toepfer@zfl-berlin.org

G. Winter
Universität Bremen, Fachbereich Rechtswissenschaft, 28353 Bremen, Germany
e-mail: gwinter@uni-bremen.de

self-regulation.[3] NGOs were quick to warn that regulation should not be left to the scientific community alone and that broader ethical and societal implications needed to be taken into account (ETC Group 2007). The debate was stirred further when some synthetic biologists used "creation of life" rhetoric to promote their work. Fear of negative reactions by the public stimulated funding of ELSI activities (Torgersen 2009). Governmental bodies and representatives of academia have commissioned reports to prevent some of the alleged mistakes and dynamics that occurred during previous genetic engineering governance (DFG et al. 2009; EGE 2009; PCSBI 2010; RCUK 2012).[4]

Publications on the potential chances and challenges often address synthetic biology in general.[5] There may be good reasons for this in some cases. For example, Deplazes et al. (2009) maintain that all fields of synthetic biology share an aim to create or recreate life, and that they therefore also share related ethical questions. Some analysis around the concept of life can thus be relevant to synthetic biology generally. However, even with regard to this particular topic it does make a difference, whether ethical implications of challenging the concept of life are analyzed in the context of "protocells" or in the context of "genetically engineered living machines". Therefore general recommendations about synthetic biology can be misleading.

## 1.2 A Mosaic-like Field on the Move

There are probably not many other scientific fields besides contemporary synthetic biology that can name their exact place and date of birth: the formal delivery room of synthetic biology is the Massachusetts Institute of Technology (Cambridge/ Mass), its birth day is the 10th June 2004, and the founding conference "SB 1.0".[6] Since then, synthetic biology has matured and spread with a remarkable speed: with its own conference series, journals, chairs and funding schemes being already

---

[3]On the conference "Synthetic Biology 2.0: The Second International Meeting on Synthetic Biology May 20–22, 2006 at UC Berkeley" a biosecurity resolution was discussed: http://synthe ticbiology.org/SB2.0/Biosecurity_resolutions.html. Accessed 18 July 2015. At the end no action on self-governance was taken.

[4]Therefore, ethicists and sociologists have soon had to reflect their roles within the emerging technology (Rabinow and Bennett 2009, 2012; Calvert and Martin 2009). This is a phenomenon known from other science and technology fields characterised by such early concomitant research (cf. nanotechnology, see Schummer 2011). Synthetic biology has also become a case study object for secondary research about policy advice.

[5]For an exception, see Rabinow and Bennett (2009).

[6]"The First International Meeting on Synthetic Biology" 10.-12.6.2004 at the Massachusetts Institute of Technology Cambridge (MIT), MA, USA, http://syntheticbiology.org/Synthetic_ Biology_1.0.html (accessed on 14 May 2013). This was when the ideas about how to revolutionize the use of open source and engineering principles in genetic engineering Endy (2005) were first presented by their protagonists. Other research lines as for example protocell research that now are commonly included under the synthetic biology umbrella term have their own roots.

right in place. It has attracted researchers not only from scientific and engineering disciplines but also philosophers and social scientists who got quickly involved in synthetic biology. But even though we know when and where it started, there is little agreement about *what exactly* started then.

To understand contemporary synthetic biology it is worth having a look at its roots. The label synthetic biology is not new and had, for instance, sporadically been used in the 1970s to describe genetic engineering. In the context of contemporary synthetic biology, however, the label 'synthetic biology' seems to have emerged without direct verbal or historical link to those earlier appearances. Moreover, tracing the disciplinary label falls short, since other names like "constructive biology or "intentional biology", had also been discussed in advance of the inaugural conference in 2004 (Campos 2009). Synthetic biology as founded at the SB 1.0 conference has developed out of several fields and contexts. Most key actors at this early time came from outside biology and were mainly specialized in computer sciences, artificial life research or engineering. They were driven by the vision of transforming biology into an engineering discipline, which was discussed in diverse scientific contexts for quite some time. But only the new achievements in molecular genetics, systems biology, next generation sequencing technologies and large-scale automated synthesis of DNA enabled synthetic biology to find the required technical preconditions and to get started. These new technological advances attracted engineers who were fascinated by the vast possibilities that living production platforms offer. They sought to make the methods of classical genetic engineering radically more effective through the application of principles already established in mechanical and electrical engineering sciences. Foundational principles like standardization and modularization were introduced, allowing for the effective design of genomes and cells (Endy 2005).

Synthetic biology thus started with the application of engineering analogies and models, that were developed for technical purposes, to biological systems. Although the field has developed and matured since then, some of the engineering analogies were difficult to realize in practice. In classical engineering practice standardized parts are utilized that work strictly independently of the chassis they are implemented in. In living systems, however, phenomena like signaling crosstalk, cell death, mutations, intercellular communication, extracellular conditions and stochastic noise have to be taken into account (Purnick and Weiss 2009). Biological systems are unique in their complexity, and synthetic biology had to integrate a lot of fundamental understanding from the biological and chemical sciences in order to get closer to achieving its goals. Although the integration of the biological and chemical sciences and engineering has probably become more appreciated (Smolke, personal communication 2012[7]), some of the leading figures

---

[7]The provided information is based on a non-public expertise conducted by Margret Engelhard and Kristin Hagen for the German Parliament. The expertise (duration: 2/2012–12/2012) is based on qualitative interviews with leading scientists from within synthetic biology, scientists that research on synthetic biology and active artists. Main content of the interviews were the current status of synthetic biology, its framing (also in comparison to genetic engineering), on xenobiology and protocell research, the individual research agendas, the role of DIY-biology and questions on potential risks.

of synthetic biology even advocate shifting back to cell free systems to manage multiple enzymes in ever more complex systems (Billerbeck et al. 2013).

In addition, researchers originating from different disciplines have rather different approaches to synthetic biology. The disciplinary background is framing the individual research agenda and also the image, aim of and approach to synthetic biology.[8] Very often the scientific socialization plays also a role in the conceptual foundations of the work. For instance, engineers might be most interested in the development of cell factories and production platforms, whereas scientists trained as chemists are more likely to focus on optimizing molecular tools or developing new genetic materials functioning as chemical alternatives to DNA (XNA, Xeno Nucleic Acids). Other lines of research that are commonly seen as part of synthetic biology include the design and construction of minimal cells and the chemical generation of protocells.

The recent discovery of a new method of genome editing called CRISPR/Cas9 (clustered regularly interspersed short palindromic repeats) has further boosted the field. This system is making use of a defense machinery of certain bacteria against viruses. Researchers figured out how to use this system to precisely cut DNA at any location on the genome and to insert new material to alter its function or to introduce new traits (Makarova et al. 2011; Jinek et al. 2012). This is a profound advantage over other techniques developed to edit the genome, such as zinc finger nucleases or TALENs. These are quite complex so only a few expert labs followed up on genome editing as a practical tool. CRISPR/Cas9 is a methodology that brings the central agenda of synthetic biology to the point. It provides a simple, standardized and highly predictable method for engineering the genomes of organisms for new purposes. A similar methodological paradigm shift has been described in the context of recombinant DNA technology as being triggered by methodical break through.

> Although revolutionary in their impact, the tools and procedures *per se* were not revolutionary. Rather, the novel ways in which they were applied was what transformed biology. (Berg and Mertz 2010).

In this sense CRISPR/Cas9 is one of the most effective tools ever developed for synthetic biology, and its potential has been by far not exploited yet.

During its short existence, synthetic biology has thus not only grown substantially and moved on at the methodological level, but also broadly expanded its scope. Part of the expansion has been achieved by active field-building initiatives: at the "SB x.0" conference series, people from neighboring disciplines were and are deliberately invited to mutually stimulate the research and possibly join the field.[9] In addition, the prominent international undergraduate student competition *iGEM* (international genetically engineered machines) attracts growing numbers

---

[8]Results interviews see FN above and Table 3.1, Chap. 3.

[9]This was stated on one strategy of the conference in the welcome address of Drew Endy at the "SB5.0: the Fifth International Meeting on Synthetic Biology" held at the Stanford University, Stanford, California USA, June 15–17, 2011.

of participants. It has already spawned several generations of students who are making their way into synthetic biology research and development.

Today, synthetic biology is a very diverse field. It comes along in a mosaic-like structure, for which there exists no sharply-cut and generally accepted definition. (Kronberger 2012, for a good overview on various definitions of 'synthetic biology' see Deplazes-Zemp 2009 "Piecing together a puzzle"). Thus, the label 'synthetic biology' serves more as an umbrella term. Nevertheless, intense identification with synthetic biology is widespread among its protagonists.[10] Thus, ethical and societal discussions have to take into account the complex and fluid state of synthetic biology. Societal evaluation needs to differentiate from case to case if it is appropriate to discuss synthetic biology as a whole, a subfield or just an exceptional mosaic stone.

## 1.3 Challenges to the Societal and Ethical Evaluation

The mosaic-like structure of synthetic biology makes social and ethical evaluation a demanding task. Consequently, the results of the ongoing societal, ethical and risk debate about synthetic biology's consequences and implications have been divergent (example reports include: BBSRC Report 2008; PCSBI Report 2010; ECNH Report 2010; Kaebnick et al. 2014 and the report of Secretariat of the CBD 2015). There is little agreement on guidelines for science, political recommendations, or prohibitions ('further discussion is needed', being the nearest statement to one of prohibition). We can find a multitude of positions, ranging from cautious and critical views (e.g. Hauskeller 2009 on protocells) to liberal laissez-faire attitudes (e.g. Eason 2012). Some authors even regard it a moral obligation to actively promote further development in synthetic biology to secure, for example, global food supply (e.g. Tait and Barker 2011).

There is also some disagreement over the importance of various issues, in particular the implications of changing concepts and definitions of life. Are the ramifications of such worldview changes central (e.g. Presidential Commission for the Study of Bioethics 2010), or are they irrelevant (e.g. Kaebnick 2009)? Will we get used to the idea that life can be synthetic, and that human beings will create life, in the same way we got used to seeing the world as a sphere (Tait 2009)?

What, then, are the ethical issues that require careful reflection? In the beginning, ethical debate within the synthetic biology community and within governmental organisations dwelt almost exclusively upon the issue of risk. And there can be no doubt that safety is of great importance: bio-safety and bio-security remain dominant issues especially within the synthetic biology community and in debates concerning government policy making (e.g. DFG et al. 2009). This almost exclusive focus did, however, stir reactions in both civil society groups and with

---

[10]Results interviews, see FN 7.

philosophers (Boldt and Müller 2008; Torgersen 2009; Friends of the Earth 2012). They asked for a more profound ethical investigation of the field. As a consequence, broader ethical issues are now increasingly being taken into account, such as problems of international justice or fundamental concerns about human beings "playing God" and re-defining "life" and "nature" (Dabrock 2009). Many of these debates have parallels in previous discussions concerning genetic engineering and cloning.

The described diverse and partly contradictory lines of debates have their origin in the multilayered structure of synthetic biology and in differences in the underlying concepts that are applied like, for example, the concept of life or artificiality. The societal, ethical and risk evaluation of synthetic biology is therefore often prone to misunderstandings and may lead to debates that go astray. To overcome the problem of the complexity of the field, we will focus on specific features of synthetic biology that might be relevant to the societal discussion rather than on synthetic biology as a whole. By this approach we hope to clarify on what aspects of synthetic biology a new societal discussion is needed. At the same time this approach will circumvent the problem of misunderstandings resulting from the fluid state of synthetic biology.

## 1.4 Relevant Features and Concepts of Synthetic Biology

Although synthetic biology builds on many of the techniques of genetic engineering, some critical features of synthetic biology are new and thus are—next to the well-known features of genetic engineering—important for an appropriate societal evaluation. Synthetic biology aims at the design and construction of new biological parts, devices and systems. Thus, it goes far beyond mere modification of existing cells by inserting or deleting single or a few genes. Instead, cells will be equipped with new functional devices or even complete biological systems will be designed. Therefore, in comparison to genetically modified organisms (GMOs), synthetic cells are characterized by a much larger **depth of intervention** into the organism. The resulting synthetic organisms may have only very little in common with any other organism, and predictions on the behavior of these synthetic cells cannot simply be drawn from the behavior of any of the many donor organisms that have been used for the creation of synthetic cells.

Another critical dimension of synthetic biology is the introduction of unnatural chemical compounds into living systems. This results in cells whose metabolism or genetic material is incompatible with that of natural biological systems. Although this **orthogonality** is sometimes discussed as a means to enhance biosafety (Marliere 2009; Budisa 2014; Budisa et al. 2014) these xeno-organisms may have an unpredictable impact on the ecosystem if released into the environment.

In addition, the more orthogonal synthetic organisms are to their natural counterparts, the more unfamiliar they are, not only for the ecosystem but also for both scientists and the public. Therefore knowledge that comes with practical experience is lacking. This unfamiliarity results in an enhanced **uncertainty** with respect to risk assessment, since any predictions are normally modelled along the known properties of related organisms we are familiar with.

Another aspect of synthetic biology becomes most apparent in those experimental set-ups where simple protocells are generated from non-living materials. Here it is the question at what point (if at all) these cell-like structures can be regarded as being alive. The claim to create live from inanimate matter will have a deep **ontological** impact on the concepts of life and artificiality.

To illustrate these dimensions of synthetic biology we introduce a layer model that analyses and classifies organisms developed by synthetic biology on an evolutionary and systemic level.

## 1.5 Layer Model

For the development of the layer model we will focus on the connectivity of synthetic or semisynthetic organisms and cells with the realm of natural organisms derived by evolution. In our model, organisms created by synthetic biology will be assigned to different "layers" on the basis of their biological and chemical nature: all organisms that use the same genetic material (e.g. DNA or any kind of xeno-DNA) **and** the same genetic code for translation of genetic information into proteins are placed within *one* layer. This implies that within such a layer any exchange of genetic material results in meaningful exchange of genetic information. With this definition, all naturally occurring organisms based on DNA as genetic material and the standard genetic code belong to one layer.

Such a description directly serves as a template to clarify a number of discussions. In the risk discussion, for example, the degree of familiarity among the organisms is essential to judge potential risks. On the basis of the proposed layer model the degree of familiarity to natural organisms can be better analyzed.

If a synthetic cell or organism differs in its chemical composition and/or uses a different genetic code it belongs to a *different* layer of organisms since no mutual exchange of genetic information is possible between such cells and those of the "natural" layer. One thread in this book will be an explication of this layer model and its application in focussing the discussion about societal and ethical consequences of synthetic biology on its new and relevant features. This approach differs from earlier descriptions of synthetic biology that are based more on categorisations of its different methodological characteristics and branches, (e.g. de Lorenzo and Danchin 2008; O'Malley et al. 2008; Schmidt et al. 2009; Deplazes 2009; IRGC 2010; Joly et al. 2011; Pei et al. 2011).

How the layer model illustrates the biological properties of natural and synthetic organisms may be exemplified for the engineering branch of synthetic biology. It aims at creating new cells by using minimal cells that will be equipped with standardized DNA modules (biobricks) conferring useful properties. Since all biobricks are currently based on the standard genetic code, synthetic cells created by this approach still belong to the layer which also contains the naturally occurring organisms. In this sense, synthetic biology can indeed be regarded as an extension of classic genetic engineering. The latter technique, however, normally generates organisms that share most of their genome with that of their natural progenitors since only a single or a few genes are exchanged. Therefore in the layer model, genetically modified organisms (GMOs) have their place in close vicinity to natural cells. In contrast, synthetic biology aspires to explore the full sequence space of all potential organisms, which have not yet been realized by evolution and thus can be termed "beyond natural" (Elowitz and Lim 2010). Therefore, synthetic cells are placed within the layer at a large distance from natural cells. This illustrates the deeper level of intervention of synthetic biology. In this regard, synthetic biology goes far beyond genetic engineering.

Since engineered synthetic cells are still able to interact with natural organisms at the genetic level, this constitutes the risk that its genetic material may spread into the environment and may cause damage to the microbial communities of the ecosystem. To prevent such unwanted exchange of genetic information it has been proposed to design cells that are **orthogonal** to natural cells. Orthogonal organisms are unable to interfere or to interact with natural ones. This can be reached, for example, by using chemical molecules that are new-to-nature as genetic storage material. If such material is released into the environment it cannot be used by other species as a useful source of information. However if other organisms cannot use such materials their degradation might become problematic and might thus lead to unwanted accumulations. Another strategy for orthogonalization would be the reassignment of one or several codons of the genetic code. In this case the chemical basis of cellular and genetic metabolism can be maintained. Exchange of genetic material (DNA) might still occur, but will not result in meaningful exchange of genetic information since the incoming DNA will be translated into a completely different protein. But even if the strategy of orthogonalization precludes unwanted spreading of genetic information into the environment one has still to consider ecological interactions at the level of competition for nutrients or toxicological aspects of the xeno-material used to construct these cells.

In contrast to the vision of most synthetic biologists to create synthetic cells by large-scale redesign of natural cells, **protocell** researchers address the problem of constructing simple life forms directly from non-living matter. With most approaches, lipid vesicles are used to mimic cellular compartments in which enzymes maintain simple metabolic networks or self-replicating polymers display basic features of growing cells (Rodbeen, van Hest 2009). As it is commonly accepted that life must have started from similar vesicles, protocell research is also a means to study prebiotic evolution by reconstructing the emergence of early life forms. Here it is the big question at which point a protocell indeed becomes a living entity. A common notion is that as long as the cell is capable of self-maintenance, reproduction, and evolvability, it can be regarded as being alive. (Luisi et al. 2006). Most of the contemporary attempts to create such cells are far from reaching this point and in no case, cells were equipped with a core translation machinery which would be necessary to express genetic information. In most cases only single aspects of self-maintenance and reproduction have been realized and often in very primitive fashion, e.g. self-catalyzed replication of a ribozyme in vitro (Wochner et al. 2011) or template-directed synthesis of a very short genetic polymer in a model protocell (Mansy et al. 2008). Therefore, these simple systems can hardly be called living and in our layer model such protocells would occupy layers of very limited size. Since RNA or DNA molecules that can be faithfully replicated by such means are only very short, the number of different sequences is very limited. This defines the very small size of a protocell layer. Some challenges of protocell biology seem insurmountable, like reinventing a translation machinery that uses genetic information to instruct the synthesis of proteins. Therefore in some approaches the plan is to circumvent this problem by implanting ribosomes derived from natural cells into protocells. Such cells are termed semisynthetic cells and might be the prime candidates to cross the border between the non-living and the living. But in the very moment where such a semisynthetic cell uses DNA and the standard genetic code it will be found on the natural layer of cells (maybe not far from minimal cells deprived of all unnecessary functions).

## 1.6 Challenge to the Concept of "Life"

With the aim of creating non-natural "life", synthetic biology obviously presupposes a general understanding of what "life" is. But, at the same time, the success of synthetic biology's project will have profound impacts on this very concept. Many protagonists in the field see the most obvious impact in a more (or possibly exclusively) mechanistic understanding of life. In the view of synthetic biologists, (artificial) living beings are composed of arranged chemical building blocks and, consequently, can be completely explained by reference to these dynamic parts and their organization. Within this conception there remains no place for vital forces or special properties of living matter, that have been evoked since antiquity in order to account for the manifest autonomy and internally driven dynamics of living beings. In our everyday notion of life these aspects of living beings are still very prominent, and they are also connected to our ethical evaluation of life as something inherently good, fragile and at the same time extremely persistent in its self-maintenance. Entirely mechanistic conceptions of life clearly challenge our well-established normative notion of life as something flourishing and striving for its own good.

In the eyes of the protagonists, the achievement of creating artificial living beings out of inanimate matter would result in the final refutation of all non-naturalistic views of life. In an almost paradoxical manner the artificiality of living beings created by synthetic biology could strengthen our naturalistic understanding of life: the possibility of making organisms synthetically by assembling non-living building blocks into an integrated organized system suggests that nature can proceed in similar ways. The only difference can be noted in the fact that in natural living systems the organizing process comes from within the system (self-organization), whereas in artificial living beings it is the result of external forces (at least at the beginning by setting the starting point for a series of self-replicating systems).

It is suggested that, with the step of synthesizing organisms, the concept of life will lose much of the mysticism which has accompanied it for centuries. By further integrating "life" into the net of naturalistic explanations, it will become part of the normal physical world. Quite adequately, then, the construction of artificial living beings can be called "one of the most important scientific achievements in the history of mankind" (Caplan 2010, p. 423). With this integration being accomplished we would no longer need to postulate a special force or power to explain life. Living beings would become nothing else but very complex molecular mechanisms.

But, on the other hand, synthetic biology, as it is already expressed in its name, is not a purely analytic and reductionist research strategy. It does not focus exclusively on the study of components but stresses the importance of higher level phenomena and their integration into the system. Therefore, synthetic biology will not necessarily strengthen a purely mechanistic view of living beings. Central concepts of the discipline like "organization", "interdependence", "mutual control", "modularity" or "redundancy" are not are not reductionist concepts. They refer to the level of the whole system and do not necessarily support the reducibility of higher level phenomena to the level of the parts.

It is therefore attractive to understand synthetic biology as a systems theoretical approach that provides concrete models for the scientific explanation of living beings in the real world. On the basis of the model of synthetic organisms, it will be possible to better understand the material and systemic implementation of the complex and holistic capacities that are characteristic of living beings and used to define life, e.g. growth, organizational closure and self-maintenance.

The three areas of synthetic biology that we propose to distinguish, i.e. engineering biology (layer of genetically modified and artificially designed cells), xenobiology (layer of orthogonal cells) and protocell research (layer of simple life forms), provide quite dissimilar models for this aim. They focus on different aspects of living beings and correspondingly their achievements will probably have different effects on the concept of life: engineering biology is especially concerned with the organizational architecture of living beings, for example the modularity of their parts; xenobiology starts with the genetic level and, at its beginning, leaves the extragenetic machinery of the cell untouched; protocell research, finally, is especially focused on the establishment of the metabolism and functional organization of a single cell. Accordingly, the synthetic projects on the three layers will provide insights into the structural design, genetic mechanism and metabolic circuits respectively. Although the insights gained by the different approaches of synthetic biology will be based on artificial systems, they will refer also to the realized systems in nature as one of the solutions among the possible.

The artificial creation of living beings will contribute to our understanding of naturally occurring organisms. At the same time, it will have impacts on the principles that scientists will use for the analysis of naturally occurring organisms. And this "creation" will also readjust the general scope of the concept of life.

This will take place in some obvious senses as the realm of the living will lose its coherence and unity given until now by the universal genealogical and ecological nexus between all living beings. By definition, the products of synthetic biology are not part of the natural genealogical nexus; and on purpose, they are (at least in many cases) designed not to interact ecologically with the naturally occurring organisms.

It is likely that this divide between two worlds within the sphere of the living will be accompanied by an increasing divergence between the scientific and the non-scientific concept of life. In its non-scientific usage, "life" is a fundamental category for the self-description of our existence; in its well established biographical sense the term refers to "the sum of one's aspirations, decisions, activities, projects, and human relationships" (Rachels 1986, p. 5). These phenomena are essentially unpredictable and have an opaque structure; they are not completely controllable but have freedom, chance and contingency as integral moments. For this reason, they seem not to be compatible with the complete mechanistic understanding of artificial organisms for which synthetic biology strives. And more than that: for reasons of safety it is exactly these aspects that synthetic biology tries to eliminate in artificial organisms. The intentionally created organisms are designed in order to be completely predictable and reliable in their behavior. To the extent that this aim will be accomplished, our scientific concept of life will change and our non-scientific concept will depart from the scientific understanding.

Another tension within the concept of life that will be established by synthetic biology refers to the teleological order and autonomy of living beings: in synthetic biology, artificial living beings are constructed with the aim of using them for our purposes, not their own: the life processes of these organisms serve not only internal but also external purposes. This seems in conflict to how we normally understand life, namely having nearly-complete functional self-referentiality. To be sure, man always has used animals and plants for his purposes. But he always had to respect their autonomy and particular way of life in order to do so. In contrast to this, synthetic biology carries the heteronomy of living beings to its extreme: the degree of the autonomy of the system is not something we are forced to work with but what we determine.

In imagination this strategy was always legitimate as the long history of the notion of "artificial life" shows, which never was a contradiction in terms. Synthetic biology forces us to take this option seriously: living beings that are designed and used for external purposes and that are completely predictable in their behavior should not lose their property of being alive.

## 1.7 Ethical Evaluation

In the film *Blues Brothers* (1980) Jake and Elwood Blues must drive 100 miles to Chicago; it is night and they are chased by hundreds of police cars. Time is tight and things do not look too good, as Elwood rightly remarks: "It's 106 miles to

Chicago, we got a full tank of gas, half a pack of cigarettes, it's dark... and we're wearing sunglasses". We might opine that they could have taken their glasses off!

When we look at synthetic biology, the situation is worse. There are certainly more than 106 miles ahead of us and our ethical vision of things to come is *structurally* limited by things we cannot simply take off like the Blue Brothers' sunglasses. It is simply too early to foresee all possible developments of synthetic biology. This limitation is a general problem with knowledge, as Karl Popper has argued: "no society can predict, scientifically, its own future states of knowledge".[11] In order to know about technological inventions of the future one would already have had to make them.

However, what we can anticipate, and in particular what we cannot, makes synthetic biology of high ethical relevance. Hans Jonas (1987) has identified five characteristics of modern technologies (for example nuclear power) that explain why these technologies pose such a problem for mankind and the environment and are thus in need of moral regulation. What are these five characteristics? Jonas first sets out the inherent ambivalence of modern technology's consequences (1987, 42f.): many of its developments show unexpected and undesired side-effects. The same is certainly true for synthetic biology, as the risk-discussions show. We might create organisms which are *prima facie* useful (e.g. in the production of bio-fuel) which could turn out to have harmful and uncontrollable effects. Secondly, Jonas points to the inherent dynamics of developing new technologies (1987, 44): fascination with the science itself, irrespective of its application, stirs scientists to do more and more research. The scientific hype around synthetic biology shows that this is very much what we find today; we see a scientific gold rush. The same is true for the third characteristic Jonas lists (1987, 45ff.), namely the spatial and temporal scale of the consequences. If synthetic biology creates artificial life, then its consequences could be far-reaching. New organisms could enter the evolutionary chain and be part of the future of life. After all, the commercial world seeks reliable, robust, and long-lasting new organisms. These might be fitter than some "traditional" life forms and could replace them according to the principles of natural selection. Jonas's fourth issue is that technology poses problems that cannot be answered within traditional anthropocentric ethics alone. In response to this problem, the last decades have seen many proposals for an environmental ethics (including Jonas's own efforts). While some of the problems raised by synthetic biology fall under the domain of traditional anthropocentric ethics (such as the problems of risks to human life), the issues certainly go beyond that. New organisms could be a danger to other organisms but we might have to decide, for example, whether we want to protect a synthetically designed species. Anthropocentric ethics will not be enough and it is still open as to whether traditional environmental ethics is sufficient. Jonas refers, finally, to a fifth characteristic (1987, 48 f.): modern technology raises fundamental ethical concerns which Jonas calls "metaphysical" issues (should, for example, the continued existence of the human

---

[11] *The Poverty of Historicism* Boston: Beacon Press (1957), p. 21.

species be thought desirable?) The risk-debates have shown how much synthetic biology adds to this concern—and radicalises some of these questions even further. What is life? What is natural? All of these questions are hard to answer, but they have become even tougher since the rise of synthetic biology, because of the intermediate status some of the developed organisms represent.

There can be no doubt that even now, at this early stage, ethics must take synthetic biology seriously. If we distinguish between the level of engineering biology, chemical synthetic biology (orthogonal life or xenolife), and proto-cell research, we can look at the five characteristics in a more careful way and anticipate possible developments more clearly. While all fundamental concerns are raised on all three levels (Jonas' fifth characteristic), the harmful and uncontrollable effects are probably mostly at the engineering, and partly on the proto-cell, level. If it is true that xenolife can eventually be isolated by a genetic fire-wall from the environment (Budisa 2014), then negative effects on this level might be minimized. However one still has to take into account effects on ecosystems of a different kind, as for example at the level of substrate competition or accumulation of toxic chemicals. The problem of how to deal with entirely new organisms will crop up mainly at the level of engineering biology and xenolife.

Even at this early stage there can be no doubt that the ethical challenges of synthetic biology are substantial—some would say dramatic—and that all developments in technology in the area certainly call for monitoring. But we do not even have adequate means for such monitoring. We have already stated that the biosafety risks of synthetic organisms are particularly difficult to assess because of their divergence from the natural and their depth of intervention. We have therefore little if any knowledge of how synthetic organisms or proto-cells might interact with ecosystems. That is why research in synthetic biology should be accompanied by work on improved methods of risk-assessment, well beyond what is typically undertaken today. (This does not mean that central underlying moral guidelines are likely to change: the central concerns remain fundamental human rights, the protection of the environment, and the demand that risks and gains should be in a reasonable proportion.)

It must be made clear that currently synthetic biology is *acting under uncertainty*, in particular in the cases of organisms that are released—wanted or unwanted—to the environment. Following Sven Ole Hansson (1996, p 369), we might even call it a "great uncertainty", because neither do we have a picture of all the options that we (or the scientists involved) have, nor do we have a clear perception of possible consequences. That is why probabilities can simply not be assigned to consequences, not even in the case of xenolife. Compared to traditional genetic engineering, this is a new quality of uncertainty, which is why a precautionary principle should rule. Highly harmful consequences are imaginable, and possible. To these we can attach no probability rating. We should avoid such risk.

Whether, in the end, a new type of ethics (an ethics of synthetic biology) is required, is too early to say. Certainly, many ethical debates in synthetic biology are simply continuations of debates we know from previous discussions regarding genetic engineering. Potential problems of international justice, for example, are discussed very in a similar way and these traditional arguments are still used in synthetic biology (new developments might solve the problems of future genera-tions—or they might deprive poor people of their resources).

The so-called "playing God" argument has sometimes been seen as unique to the ethical debate on synthetic biology. It is, however, not obvious at all that the "playing God" argument is an argument at all, or much of an intuition (Müller 2004). What could its rational core be? There are three different ways in which the playing God argument can be interpreted.

1. It might be a warning that synthetic biologists (or a society which promotes such research) overestimate their own abilities; they may be tempted to create life-forms without sufficient care, driven simply by the fascination of the pos-sible. We might call this the "hubris-interpretation".
2. Another analysis is that the discomfort arises from the limits that are trans-gressed by synthetic biology. Does artificial life violate a hitherto-unrecognized moral norm? Perhaps "life" is a limit that science should respect under all cir-cumstances? This is the "taboo- interpretation" of the argument.
3. A third analysis is to understand the objections in the spirit of Goethe's "Sorcerer's Apprentice", the pupil of a magician who exercises his master's tricks but cannot stop what he has started. Artificial organisms might develop a dynamic we cannot control. In this interpretation, the argument shares the problem of acting under uncertainty: we cannot foresee what will happen and do not have the means to recapture organisms once deliberately or accidentally released.

All these warnings are, of course, familiar from other contexts: human hubris is much discussed in political ethics, life as a possible absolute limit turns up in some bioethical debates (for example on euthanasia), and uncontrollable effects are cen-tral to environmental ethics (for example the debate on nuclear power). At least if one defines "new ethics" as an attempt to introduce *new norms or values* (as envi-ronmental ethics argued in the 20th century), then synthetic biology is not likely to require a "new" ethics. The ethical concerns it raises, such as risk, acting under uncertainty, global justice, and hubris, are familiar. But still, synthetic biology is adding a new dimension to these concerns and certainly requires close monitoring and careful attention. If "new ethics" is understood as a demand to intensify reflec-tion and debate in a *new field*, then we can certainly talk about a necessary new ethics of synthetic biology. After all, there are features of synthetic biology which require normative answers beyond the scope of traditional ethical debates.

This, at least, seems clear: synthetic biology demands careful investigation in its own right. We must understand what exactly is new and where precisely the challenges of this new technology are to be found. This can be done most effi-ciently by identifying characteristic differences between synthetic biology and genetic engineering in terms of operative features and the nature of produced objects by looking, for example, at the three levels separately.

But further differentiation will be necessary. Even if traditional normative cate-gories suffice for an ethics of synthetic biology, their exact meaning and reference is challenged. An example is the term "species". Environmental ethics generally sees a good in the protection of all species and the diversity of life forms—but does this include artificial organisms? We must also ask what exactly the 'preservation of a species' requires if extinct species can be rebuilt where we have the genetic information (here conceptual problems touch ontological ones). Another example is "life", as already discussed above. If its meaning changes we might have to reconsider basic categories. If "life" is naturalized further (as synthetic biology seems to promise), is there any reason left to treat living things differently from inanimate matter? And how do we respond if the concepts of life in a scientific and ordinary-language context differ dramatically? The reach of these conceptual changes might be profound enough to touch meta-ethical issues and problems of justification. Why should we treat organisms with special care if they do not have the functional self-referentiality of traditional organisms and if they are fully predictable in their behavior? Synthetic biology certainly challenges "old" ethics in many ways. And thus a new and profound ethical debate is required and that might change old ethics fundamentally. Synthetic biology has the power to undermine much of our traditional worldview, possibly even beyond ethics.

## 1.8 Synthetic Biology as an Object of Public Perceptions

The hallmarks of synthetic biology as an object of public perceptions are right now fuzziness and a low degree of salience. The lively societal debates accompanying the field from the outset have been confined to analysts and experts from ethics and social sciences, while the public at large remains virtually unaware of this emergent field and its many promises and challenges. Certainly, the technical complexity of synthetic biology, including the mosaic-like structure described above, poses difficulties even for scientists from other disciplines, and makes it all but impenetrable for individuals that lack even a modest level of familiarity with biology and genetics; the case in most adults, even in highly developed societies. But, at present, the main barrier is not lack of knowledge about synthetic biology, but its exiguous impact in the public domain due to low media coverage. The public's space of contact with science in the making is for the most part vicarious, relying mainly on the breadth and depth of attention paid by the mass media, which in most countries has been fairly limited (in terms of quantity, recurrence and highly conspicuous attributes).[12] Usually, for scientific breakthroughs to attain high media salience requires particularly exciting practical benefits (readily available or anticipated) or, at the other extreme, serious mishaps or accidents, vocal and publicly aired disagreements among scientists, clashes between the scientific community and policymakers and regulators, or significant campaigning in favor or against particular

---

[12]For reviews and recent data, see Ancillotti and Eriksson (in production 2016, about media coverage in Italy and Sweden) and Seitz (in production 2016, about media coverage in Germany).

developments by influential interest groups and organizations (as was the case, very early on, with plant biotechnology). The fairly scant influence of these factors, or even their absence, explains the marked contrast between, on the one hand, the profusion of reports (some by officially appointed bodies) and the boom in the ethical, social and legal literature and, on the other, the silence of an oblivious public.

Although lack of public salience, at the time of this writing, severely limits the kind of public perception studies that can be conducted, it does not completely rule out the feasibility and value of such analyses. However, the route followed must perforce be an indirect one. On the basis of findings about public attitudes toward similar cases of scientific and technological developments during their initial development stage, it is possible to anticipate the main evaluative angles that the public will apply when the field reaches a threshold of media coverage and starts to receive significantly increased attention from organized groups like environmental organizations. Biotechnology, without doubt, is the most commonly used case as the background analytical framework.

The early public perceptions literature on synthetic biology and available data from surveys and focus groups confirm that most people know very little or nothing at all about this emergent scientific area. They also document, however, that a majority (of individuals) is able to form and express views about it, falling back on very general frameworks (worldviews), more focused frames and, also, specific "cues", shortcuts, images and symbols taken from medical developments and biotechnology. Among the most relevant attributes people tend to attach to "synthetic biology", once they are briefed about its basic coordinates, are the ideas of "man-made", "artificial", "fake" and "unnatural", the trade-off between means (degree of acceptability of risks, clashes, alignment or, at a minimum, compatibility with shared values at a particular point in time and in a given society) and goals (ranking of potential benefits, from trivial to critical, fairness in the distribution of potential gains and losses). The case of biotechnology shows that specificity counts: the valence (positive, neutral, negative) of the public's response to each application is, at least partially, a function of the potential goals to be reached ("promises") and the particular nature of the means to be employed (Pardo et al. 2009). As in the case of other technological areas, medical and energy applications are in general favorably perceived, although medical ones may also generate ambivalence due to reservations regarding the means deployed (i.e., goals such as eradicating genetic diseases or new approaches to cure cancer are approved, but the use of "engineered" organisms in the human body triggers resistance).

According to Eleonore Pauwels' systematic review of the literature on public perceptions of synthetic biology published up to 2009 and the data available to 2013, the perceived risks of synthetic biology were associated to three components: first of all, the difficulty of managing unknowns, second, human and, particularly, environmental side effects, and, finally, long-term effects. These results are consistent with the findings from a major international survey, the 2010 Eurobarometer on the life sciences (Gaskell et al. 2010). From a list of seven potential aspects of synthetic biology people might wish to know more about to evaluate this new area, the majority chose the potential risks, followed by the expected benefits and the distribution of gains and risks. Interestingly, "what is being done to deal with the social and ethical issues involved" did not seem particularly salient or relevant for most people at the time the survey was conducted. Under conditions of a low level of awareness and information about synthetic biology, a relative majority, according to the Eurobarometer data, would be inclined to approve it, provided appropriate regulations were introduced, although another large group would demand stricter constraints. Around a fifth had no position on the benefits and acceptability of synthetic biology and a small, but significant minority would favor a ban on this area of research and its applications.

In terms of the governance of synthetic biology, two main dimensions stand out as relevant. The first is that the present lack of strong opposition to the field is compatible with a clear preference for tight regulation by government versus the option of giving the scientific community and corporations freedom to operate in the market, as they would in conventional business and technological areas. A second dimension of governance is the question of which group(s) provide decisive input to the regulatory process and on what basis. According to the 2010 Eurobarometer data, more than half of the European public favored scientific evidence (52 %), while just a third gave more weight to moral criteria (34 %). In line with this view, almost 6 out of 10 Europeans would prefer to rely on the input of experts in order to reach a decision about synthetic biology, rather than on the views of the majority of people in the country (Gaskell et al. 2010). These results appear to favor institutional arrangements for the regulation of synthetic biology, such as risk and bioethics committees integrated by experts, over mechanisms of social participation such as consensus conferences or, more amply, direct consultation of the public at large. However, as in other cases that are culturally shaped, there is large variability and pluralism in these preferences, with countries like Germany, Denmark and Austria giving significant weight to moral criteria and a direct public "voice", while in countries like Hungary, Spain and Italy, among others, most people are willing to rely on professional input based on scientific reasons and evidence.

As discussed above, potential developments of synthetic biology could have profound impacts on the concept of life, to the extent of blurring the borders of the evolutionary—and natural—realm (see the layer model presented in Sect. 1.5 above). Although the deeply embedded cultural dimension of synthetic biology has not yet emerged into the public sphere, assuming that it will surface once the field develops, it is plausible to anticipate a significant civic debate about the views of life entailed by the field, the role of humans, and scientists in particular, in the modification of "natural" processes, the creation of life, and even the radical redefinition of the boundaries of such a highly symbolic concept as life, so central to the social mindscape. In the absence of highly visible and differentiated applications, a global evaluative and attitudinal position on synthetic biology could develop if leading researchers in the field insist on offering symbolic narratives (such as the creation of synthetic life in the laboratory, "regenesis", living machines) that stand at odds with current worldviews and values.

If in the next few years synthetic biology impacts on the worldviews and core values of contemporary culture, its institutional or governance dimension (including diverse forms of public participation) will likely become a major issue. Today many scientists and institutions no longer dismiss the concerns of the public on the basis of their low level of scientific knowledge, knowing that a vast array of values and central worldviews, besides knowledge and information, play a role in shaping attitudes to new scientific developments. As the legal chapter in this book makes clear, citizens' cultural concerns may very well be considered a legitimate reason in the not so distant future for imposing regulatory constraints on certain subsets of science and technology. And as in other cases where culture is important as an explanatory variable for attitudes, we should expect to find significant national disparities in citizenry's response and the regulatory framework.

The case of the social reception of biotechnology shows that a vast array of variables other than scientific literacy are at work in people's attitudes to new, potentially disruptive scientific developments, the most significant being risk perceptions, trade-offs between goals and means, ethical views, and trust in science and regulatory institutions. The scientific community not only needs to do better in communicating the basic profile of its work to society at large, but should also take into account these other value-laden factors when engaging with the public, from risk considerations to ethical aspects, and a fair balance—highly specific to each subset of synthetic biology—between goals and means. It should also recognize that the just-mentioned canonical variables operate within a more general framework, the overarching culture of society, integrated by potent symbols (such as beliefs about life) and worldviews (such as images of nature). Accordingly, the cultural, implicit and open, component of its work should also be part of its "map". The worldview that affects many subsets of the life sciences is without doubt the current vision of nature and natural processes: most people are aware of the collisions between economic growth based on intensive use of science and technology and the conservation of the natural environment, and there is acute sensitivity to scientific advances that further challenge the boundaries of "natural" processes and objects. Although this component is currently dormant apropos synthetic biology, due to the low salience of the field in the media and public sphere, it may become one of the most difficult concerns to address and, for that reason, should be part of the mindset of researchers and regulators.

## 1.9 Legal Challenges

Law in modern societies approaches new scientific and technological developments in three characteristic ways: It provides forms which enable innovation and at the same time sets limits in order to prevent unwished side-effects; it provides incentives that foster research and development (R&D) development and steer it into certain directions; and it sees to that benefits from R&D results are fairly shared. Synthetic biology is also exposed to this legal interference.

20    M. Engelhard et al.

## 1.9.1 Law Enabling and Regulating Synthetic Biology

The current dynamic evolution of synthetic biology research is enabled by the constitutional freedoms of scientific research as well as of free enterprise and property. These rights allow for regulatory restrictions in the public interest. Such regulation must be introduced if constitutional rights (such as of health) of third persons or constitutional obligations of environmental protection so require.

In view of the fundamental change in socio-economic life that must be expected from synthetic biology, it would be politically wise to develop a legal basis for it. This legal basis should take a fresh look at the entire biotechnology and establish a framework not only for synthetic biology but also for genetic engineering and high tech breeding of plants and animals. That would also give justice to new developments like the CRISPER/Cas9 technology that brings the current legal framework to its limits. Constitutional law can be interpreted to enable and even require such legal framework, because biotechnology affects fundamental rights and obligations, and because it encroaches on the separation of powers which assumes that "essential" innovations should be deliberated by parliaments rather than be left to the executives.

Alternatively, the regulation of synthetic biology could also be based on the current legislation on genetically modified organisms. However, although some strands of synthetic biology falls under the definition of genetic modification others are not. They should be included in the scope of application of GMO (genetically modified organisms) laws.

More precisely, the following synthetic biology techniques are covered by the legal definition of GMOs:

- Techniques of the first strand of synthetic biology, here called radical genetic engineering, unless they create a completely new organism; for instance, the partial replacement of the content of a cell with synthesized material of natural and artificial design would still be regarded as the production of a GMO because an original cell (and its capacity to live) was used.
- Xenobiology insofar as it incorporates genetic xeno-material into a genome.

In contrast, the following synthetic biology activities are *not captured by the GMO regime*:

- Within the first strand:
  - the full synthesis of an organism and the complete replacement of the genetic material of an organism, be it of known or new design, including xenomolecules
  - the synthesis and placing on the market of bioparts

- The second strand, i.e. the designing and synthesis of a protocell or minimal cell (be it constructed top-down in the case that no additional genes are added or bottom-up)

- Within the third strand: xenobiology, insofar as it alters chemical derivatives (amino acids, proteins) of an organism thus producing a CMO (chemically modified organism[13])

Even if the non-captured kinds of synthetic biology were subsumed to the GMO regime the current methodology of risk assessment, which was developed for the genetic modification of organisms, must be examined as to its fit to synthetic biology. Considering the increasing depth of intervention and artificiality of constructions the familiarity principle used in GMO risk assessment must be replaced by a methodology which better accounts for new designs.

Considering the impossibility of proving a zero risk to health or the environment more criteria of risk evaluation should be applied. If a risk assessment concludes that the risk is low, the project should nevertheless be asked to prove that it provides a benefit to society. Moreover, cultural concerns of populations about the benefits, risks and ethics of synthetic biology should be taken seriously in the regulatory discourse and not regarded as irrational and therefore negligible. They should, if they can be substantiated, be accepted as legitimate ground for regulatory restrictions.

For the time being, until the regulatory regime will be established a moratorium should be applied on any release into the environment of organisms resulting from synthetic biology.

## 1.9.2  Law Fostering Synthetic Biology

Synthetic biology is legally incited by two major instruments: research funding and the provision of intellectual property rights.

Concerning research funding, a mechanism should be established that weighs conflicting interests and identifies justifiable goals in order to avoid public spending following the rhetorics of hypes, unfounded promises and hasty international competition. Given the wide uncertainty concerning risks, research on safety should be supported in parallel with R&D on the construction and performance of new organisms.

Intellectual property discourse and legislation have since a long time disapproved of the argument that natural phenomena should not be appropriated by private persons. It was acknowledged that discoveries and technical advancements of natural phenomena shall be patentable. This privatization of nature is now being extended to the results and methods of synthetic biology. Although this technology is still very much dependent on natural processes, material and information, the privatization appears to have more justification because of the increasing proportion of technical input. However, this proportion is still poor in relation to

---

[13]A definition of CMO is provided in Budisa (2014).

the "higher workmanship of nature" (Darwin). Life and the natural processes of its evolution are a common good which should not be appropriated by anyone. Moreover, the privatization of life forms has more and more hindered pluralistic R&D. There is a risk that the direction of R&D will gradually be determined by commercial objectives, and that they will hinder basic research, research on risks and research on benefits for society as a whole. For this reason open source systems have emerged. They deserve to be supported. In addition, the patenting preconditions should be restrictively interpreted so that patents are not provided for information accruing in the early stages of the R&D processes but only at the end when products are ready to be brought on the market.

### 1.9.3 Law Ensuring Benefit Sharing

Synthetic biology is only slowly taking notice of the fact that it will be affected by an emerging regime introduced by the United Nation's Convention on Biological Diversity (CBD) and specified by the Nagoya Protocol which acknowledges sovereign rights of states to regulate access to their genetic resources and ask for a share in the benefits drawn from R&D on them. This appropriation of genetic resources by resource states is a response of developing states to intellectual property rights (IPR) strategies of industrialized states. This is understandable although it contradicts the idea of common goods. The regime suggests a bilateral exchange, "access to genetic resources for a share in commercial revenue from R&D". This may work if the R&D process is short. It is problematic, if the valorization chain is long because it is then hardly possible to trace benefits back to a genetic resource that was sampled in a specific provider state. Therefore, common good concepts should be tried. They would consist of cooperation between users and providers of genetic resources in R&D activities and subsequent commercialization. This option is preferable because the provider state, by participating in the whole process, stays informed about its incurring revenues and—even more importantly—is able to build up its own R&D capacity.

## 1.10 Outlook

To approach synthetic biology the classical methods of ELSI fall short. Synthetic biology is a very heterogeneous and dynamic field surrounded by grandiose visions on its scope and its economical and societal impact. These visions entail deep dimensions and concepts like that of life and our ability to reshape nature and evolution radically. Transferring the full set of engineering principles and visions from the material world to living beings has a deeper impact on both culture and nature than canonical biotechnology using genetically modified organisms (GMOs). Without taking this dimension into account, many discussions on synthetic biology appear rather general and therefore misleading; Synthetic biology is rooted not only in genetic engineering but also in mechanical and electrical engineering, organic chemistry, theoretical physics and computer science to name but a few. Thus, it has elements of both continuity and novelty. We suggest that the discussion of synthetic biology should be focussed more on the specific features that it brings up.

The aims of this book are therefore

- to identify the new features of synthetic biology in a conceptual analysis
- to frame the relevant features in relation to a categorization of research lines and envisioned applications,
- to apply these conceptual tools in the societal, ethical and legal analysis and judgment of specific features of synthetic biology,
- and, on the basis of this differentiated analysis, to suggest structures for the societal, ethical and legal discussion and policy recommendations.

This book addresses not only policy makers and interested members of the public, but specifically also the scientific communities. It is an appeal to scientists that one of their main responsibilities towards society is to be open and realistic, and also—depending on the case—to allow the critical question "is it worth it?"

# References

Acevedo-Rocha CG, Budisa N (2011) On the road towards chemically modified organisms endowed with a genetic firewall. Angew Chem Int Ed Engl 50:6960–6962

BBSRC Report (2008) Balmer A, Martin P 2008. Synthetic biology. Social and ethical challenges. Institute for Science and Society. University of Nottingham, <http://www.synbiosafe. eu/uploads///pdf/synthetic_biology_social_ethical_challenges.pdf, Accessed 18. July 2015

Berg P, Mertz JE (2010) Personal reflections on the origins and emergence of recombinant DNA technology. Genetics 184:9–17

Billerbeck S, Härle J, Panke S (2013) The good of two worlds: increasing complexity in cell-free systems. Curr Opin Biotechnol. [Epub ahead of print]

Boldt J, Müller O (2008) Newtons of the leaves of grass. Nat Biotechnol 26:387–389

Budisa N (2014) Xenobiology, New-to-Nature Synthetic Cells and Genetic Firewall. Curr Org Chem 18:936–943

Budisa N, Kubyshkin V, Schulze-Makuch D (2014) Fluorine-rich planetary environments as possible habitats for life. Life: open access J 4(3):374–385

Calvert J, Martin P (2009) The role of social scientists in synthetic biology. EMBO Rep 10:201–204

Campos, L. (2009) That was the synthetic biology that was. In: Schmidt M, Kelle A, Ganguli-Mitra, A, de Vriend H (eds), Synthetic biology the technoscience and its societal consequences. Springer, New York, pp 5–21

Caplan A (2010) The end of vitalism. Nature 465:423

CBD, secretariat of the convention on biological diversity (2015). Synthetic biology, Montreal, Technical series no. 82, p 118

Cho MK, Magnus D, Caplan AL, McGee D and the Ethics of Genomics Group (1999) Ethical considerations in synthesizing a minimal genome, Science, 289:2087–2090

Dabrock P (2009) Playing God? Synthetic biology as a theological and ethical challenge. Syst Synth Biol 3:47–54

de Lorenzo V, Danchin A (2008) Synthetic biology: discovering new worlds and new words. EMBO Rep 9:822–827

Deplazes A (2009) Piecing together a puzzle. An exposition of synthetic biology. EMBO Rep 10:428–432. doi:10.1038/embor.2009.76

DFG (Deutsche Forschungsgemeinschaft), acatech (Deutsche Akademie der Technikwissenschaften), Leopoldina (Nationale Akademie der Wissenschaften) (2009) Synthetische Biologie/Synthetic biology. <http://www.dfg.de/download/pdf/dfg_im_profil/reden_stellungnahmen/2009/stellungna hme_synthetische_biologie.pdf>. Accessed 18 July 2015

Eason RE (2012) Synthetic biology already has a model to follow. Ethics, Policy and Environ 15(1):21–24

ECNH Report (2010) Report of the federal ethics committee on non-human Biotechnology. Synthetic biology—ethical considerations. www.ekah.admin.ch. Accessed 14 Nov 2014

Elowitz M, Lim WA (2010) Build life to understand it. Nature 468:889–890

Endy D (2005) Foundations for engineering biology. Nature 438:449–453

European group on ethics in science and new technologies to the European commission (EGE) (2009) Ethics of synthetic biology: opinion no 25, European commission, Luxemburg. <http://ec. europa.eu/bepa/european-group-ethics/docs/opinion25_en.pdf. Accessed 22 Apr 2013

Friends of the earth, CTA and ECT (2012) The principles for the oversight of synthetic biology, <http://www.foe.org/news/blog/2012-03-global-coalition-calls-oversight-synthetic-biology Accessed 19. July. 2015

ETC group (2007) Extreme genetic engineering: an introduction to synthetic biology. <http:// www.etcgroup.org/sites/www.etcgroup.org/files/publication/602/01/synbioreportweb.pdf. Accessed 22 Apr 2013

Gaskell G, Stares S, Allansdottir A, Allum N, Castro P, Esmer Y, Fischler C, Jackson J, Kronberger N, Hampel J, Mejlgaard N, Quintanilha A, Rammer A, Revuelta G, Stoneman P, Torgersen H, Wagner W (2010) Europeans and biotechnology in 2010 winds of change? Technical report. European commission, brussels. <https://ec.europa.eu/research/swafs/pdf/ pub_archive/europeans-biotechnology-in-2010_en.pdf Accessed 19. July 2015

Hansson SO (1996) Decision making under great uncertainty. Philos Soc Sci 26:369–386

Hauskeller C (2009) Toward a critical evaluation of protocell research In: Bedau M, Parke E (eds) The ethics of protocells. moral and social implications of creating life in the laboratory. MIT Press, Cambridge, pp 590–641

IRGC (International risk governance council) (2010) Guidelines for the appropriate risk governance of synthetic biology < http://www.irgc.org/IMG/pdf/irgc_SB_final_07jan_web.pdf>. Accessed 19 July 2015

Jinek M, Chylinski K, Fonfara I, Hauer M, Doudna JA, Charpentier E (2012) A programmable dual-RNA—guided DNA endonuclease in adaptive bacterial immunity. Science 337:816–821

Joly PB, Laurent B, Marris C, Robinson D (2011) Biologie de synthèse: conditions d'un dialogue avec la société. Etude pour le Ministère de l'Enseignement Supérieur et de la Recherche (Sciences et Société)(Convention n11 G 603). <http://www.kcl.ac.uk/sspp/ departments/sshm/research/csynbi-PDFs/EtudebiosynthrapportFinal.pdf > . Accessed 19 July 2015

Jonas H (1987) Technik, Medizin und Ethik. Suhrkamp, Frankfurt

Kaebnick GE (2009) Should moral objections to synthetic biology affect public policy? Nat Biotechnol 27:1106–1108

Kaebnick GE, Gusmano MK, and Murray TH (2014) The Ethics of Synthetic Biology: Next Steps and Prior Questions, Synthetic Future: Can We Create What We Want Out of Synthetic Biology?, special report, Hastings Center Report 44, no. 6

Kronberger N (2012) Synthetic biology: taking a look at a field in the making. Public Understanding of Science 21:130–133

Luisi PL, Ferri F, Stano P (2006) Approaches to semi-synthetic minimal cells: a review. Naturwissenschaften 93:1–13

Makarova KS, Haft DH, Barrangou R, Brouns SJJ, Charpentier E, Horvath P et al (2011) Evolution and classification of the CRISPR-Cas systems. Nat Rev Microbiol 9:467–477

Mansy SS, Schrum JP, Krishnamurthy M, Tobé S, Treco DA, Szostak JW (2008) Template-directed synthesis of a genetic polymer in a model protocell. Nature 454:122–125

Marliere P (2009) The farther, the safer: a manifesto for securely navigating synthetic species away from the old living world. Syst Synth Biol 3:77–84

Müller A (2004) Lasst uns Menschen machen!, Stuttgart Kohlhammer

O'Malley MA, Powell A, Davies JF, Calvert J (2008) Knowledge-making distinctions in synthetic biology. BioEssays 20:57–65

Pardo R, Engelhard M, Hagen K, Jørgensen RB, Rehbinder E, Schnieke A et al (2009) The role of means and goals in technology acceptance. A differentiated landscape of public perceptions of pharming. EMBO Rep 10:1069–1075

Pauwels E (2009) Review of quantitative and qualitative studies on US public perceptions of synthetic biology. Syst Synth Biol 3:37–46

PCSBI (Presidential commission for the study of bioethical issues) (2010) New directions: the ethics of synthetic biology and emerging technologies. PCSBI, Washington, D.C

Pei L, Schmidt M, Wei W (2011) Synthetic biology: An emerging research field in China. Biotechnol Adv 29:804–814

Purnick PEM, Weiss R (2009) The second wave of synthetic biology: from modules to systems. Nat Rev Mol Cell Biol 10:410–422

Rabinow P, Bennett G (2009) Synthetic biology: ethical ramifications 2009. Syst Synth Biol 3:99–108

Rabinow P, Bennett G (2012) designing human practices: an experiment with synthetic biology, Chicago

Rachels J (1986) The end of life. Euthanasia and morality, Oxford

RCUK (Research councils UK), UK synthetic biology roadmap coordination group (2012) A synthetic biology roadmap for the UK. Technology strategy board, Swindon. <http://www.rcuk.ac.uk/Publications/reports/Pages/SyntheticBiologyRoadmap.aspx. Accessed 25.04.2013

Rodbeen R, van Hest JCM (2009) Synthetic cells and organelles: compartmentalization strategies. BioEssays 31:1299–1308

Schmidt M, Ganguli-Mitra A, Torgersen H, Kelle A, Deplazes A, Biller-Andorno N (2009) A priority paper for the societal and ethical aspects of synthetic biology. Syst Synth Biol 3:3–7. doi:10.1007/s11693-009-9034-7

Schummer J (2011) Das Gotteshandwerk. Die künstliche Herstellung von Leben im Labor, Berlin

Tait J (2009) Upstream engagement and the governance of science. EMBO Rep 10:S18–S22. doi:10.1038/embor.2009.138

Tait J, Barker G (2011) Global food security and the governance of modern biotechnologies. EMBO Rep 12:763–768. doi:10.1038/embor.2011.135

Torgersen H (2009) Synthetic biology in society: learning from past experience? Syst Synth Biol 3:9–17. doi:10.1007/s11693-009-9030-y

Wochner A, Attwater J, Holliger P (2011) Ribozyme-catalyzed transcription of an active ribozyme. Science 332:209–212

# Chapter 2
# Synthetic Biology: Diverse Layers of Live

Michael Bölker, Margret Engelhard and Nediljko Budisa

**Abstract** Prerequisite for any evaluation of synthetic biology is the precise description of its scientific rationale and its biological objects. Here, we develop a layer model that helps to categorize subfields of synthetic biology along their operative procedures and based on the biological status of the organisms generated by synthetic biology. The layer model classifies synthetic and semisynthetic organisms and cells according to their genetic connectivity and to their potential interaction with natural organisms derived by evolution. We use the model to characterize three distinct approaches within synthetic biology: engineering biology, xenobiology and protocell research. While the latter approach generates organisms that hardly could be termed living, xenobiology aims at orthogonal living systems that are disconnected from nature. Synthetic engineering biology could be considered as extreme form of gene technology since all resulting organisms share the universal genetic code with the natural living beings and are based on the same molecular and biochemical principles. Such biological description can be used to determine both the degree of familiarity and the level of uncertainty associated with synthetic organisms and may thus facilitate to judge potential risks of synthetic biology.

M. Bölker (✉)
Fachbereich Biologie, Philipps-Universität Marburg, Karl-von-Frisch-Str. 8,
35032 Marburg, Germany
e-mail: boelker@uni-marburg.de

M. Engelhard
EA European Academy of Technology and Innovation Assessment GmbH,
Wilhelmstr. 56, 53474 Bad Neuenahr-Ahrweiler, Germany
e-mail: engelhar@hs-mittweida.de

N. Budisa
Department of Chemistry, TU Berlin, Müller-Breslau-Straße 10,
10623 Berlin, Germany
e-mail: nediljko.budisa@tu-berlin.de

© Springer International Publishing Switzerland 2016
M. Engelhard (ed.), *Synthetic Biology Analysed*, Ethics of Science
and Technology Assessment 44, DOI 10.1007/978-3-319-25145-5_2

## 2.1 Introduction

Any appraisal of synthetic biology should distinguish between the narrative of its naming and its conceptual and experimental development. Synthetic biology has to be seen in the light of the different scientific traditions that have contributed to its formation, in particular gene technology, organic chemistry, chemical engineering and computer science. The term "Synthetic Biology" has been first used by the French chemist Stéphane Leduc in 1912 in his treatise "La Biologie Synthéthique" (Leduc 1912). In his experiments he studied the growth of chemical salt crystals mimicking the appearance of biological processes in his "Jardins Chimiques". He compared the osmotic growth of these forms to that of living beings and argued that in studying these formations one can learn about basic processes of life. His contemporaries, however, criticized that his growing crystals can be regarded only as an imitation of life but not as synthetic life (Campos 2009). The term "Synthetic Biology" was reinvented in the 1970 by the Polish-American geneticist W. Szybalski to describe the newly developed techniques of recombinant DNA technology: "The work on restriction nucleases not only permits us easily to construct recombinant DNA molecules and to analyze individual genes, but also has led us into the new era of synthetic biology where not only existing genes are described and analyzed but also new gene arrangements can be constructed and evaluated" (Szybalski and Skalka 1978). These historical records could suggest a connection between genetic engineering and the naming of contemporary synthetic biology. Though Luis Campos could demonstrate that the inventors of modern synthetic biology did neither know about Leducs early publications nor the further use of this term (Campos 2009). During the preparation of the first international conference on synthetic biology the organizers even discussed the alternative name "Intentional Biology" for this emerging branch of biology. This designation should emphasize the purposeful and applied nature of an engineering biology. The name "Intentional Biology" was, however, promptly abandoned since it could imply that other fields of biology would be "unintentional". The term "Synthetic Biology" was then picked to stress the analogy with "Synthetic Chemistry" (Campos 2009). Against this background, the notion of a synthetic biology refers to the synthetic approach that revolutionized organic chemistry in the 19th century. It certainly should also evoke the huge economic impact of synthetic organic chemistry on the development of the European dye and pharmaceutical industry (Yeh and Lim 2007).

Clearly, synthetic biology would not have been conceivable without the previous and successful development of genetic engineering. It was the elucidation of the universal nature of the genetic code that paved the way for experimental transfer of genes between unrelated organisms (Jackson et al. 1972; Cohen et al. 1972, 1973). These experiments are considered as genuine experimental breakthrough and mark the starting point of modern biotechnology. As early as in 1977, Genentech, one of the first biotech companies, reported the production of a human protein manufactured in bacteria: Somatostatin, a human growth hormone-releasing inhibitory factor. Since then, genetic engineering has been used in a vast number of applications, from the production of useful molecules, such as biopharmaceuticals, the sustainable generation of renewable fuels, to the design of transgenic plants that are resistant against pathogens and herbivores. Characteristic for this classic period of gene technology was the transfer of single genes or of small clusters of genes into recipient organisms, predominantly the bacterium *Escherichia coli*. Alternatively, single genes were mutated in bacteria, fungi, animals or plants to abolish or change their function. The basic concept of genetic engineering, however, did not aim primarily at reconstructing or improving nature as such. Rather, particular genetic elements were recombined or targeted mutations were introduced into genes to improve certain properties or to enhance productivity. Nevertheless, the experimental foundations of current synthetic biology resulted from that approach when the first conceptual ideas for a radical improvement of nature were developed (Szybalski and Skalka 1978). But realization of these early concepts had to wait for major new developments in the fields of systems biology and bioinformatics that together with the rapid technological progress in DNA analysis and synthesis finally allowed sequencing of whole genomes and large-scale gene synthesis.

Although gene technology and synthetic biology share many common roots, the latter has to be viewed in the light of a larger diversity of scientific traditions such as synthetic organic chemistry and, as described in the following chapters, also in the light of modern engineering and bioinformatics. In addition, the research agenda of synthetic biology clearly surpasses that of gene technology. Most important is the change of focus from manipulation to construction. It is not single genes or gene clusters any more that are mutated or recombined, but complete biological systems that are reconstructed from scratch or redesigned as a whole (Schwille 2011; Bohannon 2011). Synthetic biology even comes along with its own artistic visions (Reardon 2011).

Another important difference between gene technology and synthetic biology is the depth of intervention exerted on the organisms, both in relation to quantity and quality. In general, organisms created by synthetic biology are much more distant to any natural beings than a genetically modified organism which still can be recognized as variant of a natural species. With its methodical approach to rely more on systemic and engineering principles, synthetic biology resembles more applied sciences than basic sciences. Furthermore, synthetic biology departs from classical biological thinking and includes elements of playfulness such as the iGEM competition with its creation of bacteria that smell of bananas and wintergreen, are able to count or to blink. Nevertheless, there is no sharp boundary between classic gene technology and synthetic biology, since many applications of synthetic biology can still be regarded as gene technology. Furthermore, also gene technology is developing and sometimes reaches levels of intervention similar to that of synthetic biology. New technologies, such as CRISPR/Cas9-mediated genome editing will in future even more obscure the conceptual differences and hamper a clear distinction between these two disciplines (Cameron et al. 2014). For many critics of synthetic biology, synthetic biology is anyway nothing else than an extension of gene technology and has been called 'extreme genetic engineering' (ETC group 2007). For now, however, it may suffice to state that synthetic biology in its core area and with its conceptual and methodical impetus clearly surpasses classic gene technology.

According to its own definition, synthetic biology deals both with the design of new biological parts, devices and systems and/or the redesign of existing, natural biological systems for useful purposes (see definition on http://syntheticbiology.org). To do so engineering principles like modularization and standardization will be implemented that allow the design and description of standard biological parts. These parts can then be used to equip minimal cells with defined functionalities. The resulting synthetic cells have no direct counterpart in nature but can be used for chemical, pharmaceutical, medical and biotechnological purposes. Another aspect of synthetic biology is the generation of protocells derived from simple and defined molecules. Here, the major goal is the reconstruction of living systems from first principles. In recent years, a third approach has become more and more important aiming at the integration of non-natural (xenobiological) compounds into complex living systems. Due to the different chemical makeup of their genetic material, such cells cannot exchange any genetic information with natural organisms. A similar isolation can also be reached by changing the genetic code through reassigning the translation of triplet codons to other amino acids. In both cases the resulting organisms cannot interact genetically with the environment and thus are called orthogonal. Importantly, orthogonality of synthetic cells provides a high level of biosafety (Marlière 2009; Schmidt 2010; Mandell et al. 2015). Furthermore, due to their chemical alterations, xenobiological approaches have the potential to greatly expand the diversity of potential life forms.

Interestingly, these divergent fields of synthetic biology differ not only in their basic experimental approaches but alike root in different academic disciplines. Many stakeholders of synthetic biology have been drawn into this field from other scientific or engineering disciplines. Therefore, they have been trained in diverse fields and this has shaped their views on synthetic biology. This becomes apparent not only in their view on synthetic biology and in their conceptualizations of life but also in their attitudes of risk and biosafety management.

## 2.2 The Main Research Agendas of Synthetic Biology

Here, we will describe the basic concepts of the major subfields of synthetic biology. In general, three main areas of synthetic biology can be described as

- **Engineering Biology**, aiming at the transformation of biology into an engineering discipline by introducing standardized modules, parts and devices with well-described characteristics that can be used to construct novel biological systems or to redesign existing living systems.
- **Orthogonal Biology**, trying to create cells that are unable to exchange genetic information with natural organism either by integrating non-natural molecular compounds (xeno-life) or by reassignment of the natural genetic code (recoded life).
- **Protocell Research,** aiming at recapitulation of prebiotic evolution by building simple cellular vesicles that fulfil at least some criteria of living systems.

## 2.2.1  The Engineering Branch of Synthetic Biology

Traditional molecular biology and gene technology at large both depend on the natural genetic material as it can be found in existing living organisms. The advent of modern DNA synthesis and next generation sequencing technologies makes it possible to read and write DNA in large scale and with high precision. This has not only greatly enlarged the repertoire of available natural genes and enzymes, but also allows to design novel polypeptides without any precedent in nature. The opportunity to use DNA and protein molecules just like any other technically malleable material has attracted engineers from different fields into Synthetic Biology. From an engineer's point of view, a living cell can be considered as highly integrated complex system consisting of biochemical modules, functional devices and higher order signalling systems that works like any other hierarchical technical system (Andrianantoandro et al. 2006). All biological functions of a living cell, like environmental sensing, motility, metabolism, information processing etc. are realized by molecular complexes that fulfil specific functions. Thus, the whole cell looks like a technical apparatus well-designed to exert a complex and purposeful function. However, all simplistic approaches to reconstruct or redesign natural cells have been hampered by the high level of complexity of living cells. Many of the functional modules display intricate interdependencies that are not simply due to their obvious function. Natural organisms are the product of evolution and thus underlie constraints both from their phylogenetic origin and ontogenetic development. In addition, the process of molecular evolution by random mutation and selection entails the organisms with historical contingency. This may explain why some synthetic biologists have a delicate relationship to evolution. They often describe the complexity of living entities that result from their evolutionary origin as unnecessary (Trafton 2011) or complain that "… the design of natural biological systems are not optimized by evolution for the purposes of human understanding and engineering" (Endy 2005). On the other hand, natural evolution is a very efficient optimization strategy and synthetic biologists have also applied the technique of accelerated evolution to reprogram bacterial cells (Wang et al. 2009).

One of the main initial tasks of synthetic biology was the transformation of biology into an engineering technology. Accordingly, one of the major foundational efforts of synthetic biology dealt with refactoring biological systems to make them better suited for engineering purposes (Chan et al. 2005; Voigt 2011). As a direct way to this aim the introduction of basic engineering principles such as modularization and standardization was proposed. The most visible outcome of this approach was the construction of standardized "biobricks" organized in a registry of standard biological parts (Knight 2003; Canton et al. 2008). Biobricks are designed to allow fast, efficient and rational design of living systems by human engineers.

A second aspect of modularization and standardization is the introduction of an abstraction hierarchy of communication, which is typical for the manufacturing of highly integrated technical systems composed from many different parts and devices (Endy 2005). Standardization of parts and interfaces relieves the engineers from understanding all details of a complex system instead they are responsible only for a certain layer of functional complexity. They design specialized modules and devices that are used by other engineers to be integrated into higher systems. While it may not be necessary to understand the function of those modules and devices in all details, it is of prime importance that the technical specifications and functional parameters of these parts are described quantitatively. Due to this abstraction the standardized parts and modules can be regarded as black boxes as long as they work according to their specifications (Endy 2005). Only with this approach, an engineering technology of highly integrated living systems would be feasible in large scale.

Another important strategy of the engineering branch of synthetic biology is the use of minimal cells as platform (or chassis) for the construction of more complex systems. Minimal cells just contain all genes and proteins necessary to provide the basic living functions. The usual way to define a minimal cell is to strip natural cells from all non-essential functions. This is normally reached by random mutagenesis in combination with identification of the core essential functions by comparative genomics (Juhas et al. 2012; Acevedo-Rochas et al. 2013; Stano and Luisi 2013). Such minimal cells can then be equipped with additional modules that confer specific functionalities.

## 2.2.2 Orthogonal Biology

The engineering branch of synthetic biology primarily aims at redesigning natural cells to make them more useful for human purposes. Thus, 'natural' molecules are used to create 'artificial cells' that perform 'unnatural functions' (Benner and Sismour 2005). A complementary approach is characteristic for the 'chemical' branch of synthetic biology. Here the aim would be to use 'unnatural' chemical compounds to reproduce biological behaviour without making an exact molecular replica of a natural living system (Benner et al. 2011). Such molecular mimicking of a biological system requires an intimate understanding of the chemical properties of complex molecules and therefore mostly chemists are involved in promoting artificial life based on an alternative molecular design (Luisi 2007). Interestingly, this approach addresses another important aspect of engineering integrated technical systems, the notion of decoupling or orthogonality. Technical systems and devices are called orthogonal if they are completely separated from each other and thus do not show any unwanted interactions or crosstalk. This makes it possible to add parts to a system without creating or propagating side effects. Thus, even large systems containing large numbers of parts and devices can be designed and actually operated successfully. In synthetic biology, orthogonalization refers primarily to the inability of artificial cells to exchange genetic material and/or metabolites with natural organisms. Orthogonality is at odds with the canonical research focus in biology, where most effort is put on the detection and unraveling of more and more intricate networks of interactions. Orthogonalization aims at reversing this view and implies that system simplification might actually be at the heart of implementing a reliable and robust bioengineering. The prime route to this simplification goes through orthogonalization of biosystems to limit unpredictable (e.g. side) reactions and interactions. Current research in synthetic biology makes a lot of efforts to demonstrate the power of orthogonalization as a biosystems engineering strategy (An and Chin 2009; Schmidt and de Lorenzo 2012).

Orthogonality of biological systems can be reached by different means and at different levels. The most extreme version would be the total construction of a living cell only from unnatural components. To stress the alien nature of these artefacts and their large distance to existing biological entities these cells are also termed 'xenobiotic'. One has to point out, however, that orthogonalization is scalable and already low-level incorporation of unnatural components into existing organisms could result in considerable genetic and biological isolation (Marlière 2009; Mandell et al. 2015). In principle, it would be sufficient to replace only one of the natural base pairs by xeno-DNA bases (XNA) to prevent genetic exchange with other organisms in the natural environment (Malyshev et al. 2014).

The universality of the genetic code is the basis for gene technology since it allows functional expression of heterologous genes in cells of different hosts. Also in the natural environment genes can be laterally transferred between species. Horizontal gene transfer is a major factor of adaptation and speciation and is assumed to have occurred at large scale during natural evolution. Thus, any artificial reassignment of the genetic code is an efficient means to prevent unwanted exchange of genetic information with the organismic world. Recoded cells are genetically isolated and even resistant to the attack of viruses since viral reproduction depends on genetic compatibility. In analogy to computer networks, the genetic isolation of orthogonal cells has been termed a 'genetic firewall' (Marlière 2009). Although not the only advantage of orthogonal cells, this property is largely exploited to justify xenobiological approaches as an ultimate safety tool for synthetic biology (Schmidt 2010; Schmidt and de Lorenzo 2012).

In 2010 the first chemical synthesis of a complete bacterial genome has been achieved (Gibson et al. 2010). Therefore, reassignment of the genetic code is in reach. Just recently, in a recoded *E. coli* strain all UAG stop codons have been replaced by synonymous UAA codons demonstrating the principal feasibility of this approach (Isaacs et al. 2011; Lajoie et al. 2013). Even sense codon reassignments should be possible at least for amino acids whose aminoacyl-tRNA synthetases do not depend on recognition of the tRNA anticodon loop. In principle, genome-wide replacement of a single defined codon together with the respective exchange of the corresponding tRNA anticodon sequence would be sufficient to reassign the genetic code. Although recoded cells would still be able to exchange genetic material with the environment, natural cells taking up this DNA cannot read the genetic information of codon-reassigned organisms.

Furthermore, recoding would also allow to expand the coding capacity of the genetic code towards additional, non-natural amino acids. Therefore, efforts to expand the genetic code are expected to become a core discipline in synthetic biology as it offers an efficient platform for the transfer of numerous chemistries from the synthetic laboratory into the biochemistry of living cells (Hoesl and Budisa 2012). Any attempt to generate sequence diversity by classical protein engineering is always limited to the set of 20 canonical amino acids. This set of building blocks does not span all dimensions of chemical variability that could be potentially advantageous to diversify the catalytic performance of enzymes or to gain desired features for whole cells. For that reason, synthetic non-canonical amino acids represent an ideal tool to supply proteins and cells with novel and unusual functions and emergent features. Thereby, successful experiments include (1) uptake/import of non-canonical amino acids (2) their intracellular accumulation at levels high enough for efficient substrate turnover (activation and tRNA acylation) by aminoacyl tRNA synthetases, (3) metabolic and chemical stability of imported non-canonical amino acid; (4) tRNA charging (acylation) that must be achieved at a reasonable rate, and (5) the translation of the non-canonical amino acid into a nascent polypeptide chain. A second approach relies on the orthogonalization of protein translation as a tool for protein engineering whereby various orthogonal-pairs were developed (An and Chin 2009; Wang et al. 2014). It enabled the incorporation of around 40 different amino acid analogs, for different application purposes such as crosslinking, site specific coupling reactions etc. (Liu and Schultz 2010).

The generation of artificial genetic systems would certainly require additional and chemically different coding units. For that reason, much work has been focused on biomimetic chemistry of the DNA bases, and recently few research teams demonstrated novel DNA bases and base pair structures capable for stable unnatural base pair formation, with the ability to serve as substrate for polymerase enzymes (Pinheiro et al. 2012; Hunter 2013). This is indeed an important step in the direction to explore the possibilities of employing a new alphabet in order to change/modulate/reprogram already existing replication, transcription and translation machineries of living cells (Malyshev et al. 2014). Here we shortly outline the pioneering research in this field as newest progress is covered in numerous recent reports publications.

As DNA forms stable double helices via weak interactions between only two base pairs (A = T and G ≡ C) and exhibits considerable resistance to hydrolytic cleavage, new genetic alphabets composed of unnatural base pairs can be synthesized by shuffling the hydrogen bonding sites in the nucleobases. It was also shown that chemically synthesized mRNA containing the modified nucleotides is able to direct position-specific incorporation of 3-iodotyrosine into a polypeptide via a suppressor tRNA containing the unnatural anticodon (Chin et al. 2003; Hammerling et al. 2014). In the meantime, many research groups succeeded to greatly expand the number of synthetic base pairs capable of transcription into mRNA. For example, in vitro transcription/translation systems where DNA with unusual base pairs can be transcribed into RNA molecules and subsequently participates in protein synthesis on the ribosome were also reported (Hirao and Kimoto 2012).

## 2.2.3 Protocell Biology—Recapitulation of Evolution

Darwins explanation for the origin of species by means of natural selection immediately provoked the question for the first appearance of life on earth. The common origin of all living species and the absence of any present-day spontaneous generation of life indicate that this event may have happened only once. Singular events, however, are difficult to treat by any scientific theory due to lack of reproducibility. Nevertheless, experiments have been designed to get clues as to how life might have been spontaneously arisen in the primordial soup. Already in the early 1950ies Miller and Urey demonstrated that common organic molecules such as amino acids and nucleobases are formed spontaneously in a chemical environment that resembled the early reducing atmosphere of the earth if energy was provided mainly in the form of electric discharges (Miller 1953). Based on theoretical considerations, Manfred Eigen proposed a model where a replicating system results from cyclic coupling of enzymatic reactions (hypercycle) (Eigen and Schuster 1977). Therefore many of the approaches in protocell research concentrate on the chemical realization of such life-like reaction networks. It is generally assumed that compartmentalization in lipid vesicles is the most likely scenario that enabled these simple replicators to become subjects of Darwinian evolution (Mansy and Szostak 2009; Loakes and Holliger 2009). There is ample evidence that at a certain stage during evolution RNA formed the material basis of information storage but was also able to catalyse chemical reactions (ribozymes) (Joyce 2002). However, how stable replication and faithful distribution of these complex molecules was accomplished in such early versions of protocells is still largely unknown.

In the present era of synthetic biology, the rational design of protocells may provide an alternative way for the generation of minimal cells that could serve as chassis for the rational design of modularized cells ('bottom-up approach'). However, the lack of any knowledge on how early life forms looked like during prebiotic evolution on earth together with the very limited capacities of simple replicating systems makes it likely that it will be still a long way to a stable protocell endued with the basic properties of a living system. But on this route, important insights into possible scenarios for prebiotic evolution will be gained. Since the formation of a self-sustaining autocatalytic chemical network is a necessary step to establish simple life-like conditions, organic chemists interested in biological evolution have developed numerous self-replicating or autocatalytic systems (Mansy and Szostak 2009; Attwater and Holliger 2014). By careful inspection of metabolic synthetic processes it is almost always possible to detect some basic principles. For example, despite their diversity the amino acid synthesis pathways in extant cells share two basic features (a) the nitrogen of the $\alpha$-amino group originates from $NH4^+$ and (b) the sources of skeletal carbons are intermediates of the tricarboxylic acid cycle and the other major metabolic pathways that are coupled to the assimilation of $CO_2$ (in autotrophs) (Pereto 2012). In general, this chemical determinism should be kept in mind when considering the possibility to create living cells with biopolymers whose structure and function tolerate a systemic chemical change in the backbone such as the substitution of one atom by another: phosphorus by arsenic, carbon by silicon or oxygen by sulphur etc. That implies that all extant cells are characterised by an invariant basic chemical organisation i.e. all the properties of living beings rest on the fundamental mechanism of molecular invariance.

In addition to terminological issues, inquiries into the ways of creating life reveal that there is no consensus among scientists, philosophers, and theologians about the term "life". Thus, the aim of "creating life" is a vague scientific goal as nobody exactly knows what life itself is (Luisi 1998). The different subfields of Synthetic Biology not only differ in their methodological approaches but also in their underlying conceptions of life. The products of synthetic biology itself are at the borderline between living and non-living matter (Deplazes and Huppenbauer 2009). Depending on the different approaches of synthetic biology, certain aspects of "being alive" are emphasized. For chemists trying to reconstruct simple life-like cellular structures, self-maintenance of metabolic reactions already suffices for a minimal definition of life. Luisi has introduced the term "autopoiesis" into protocell research (Luisi 2003). A metabolic process enclosed in a semipermeable spherically closed membrane is called a minimal autopoietic system, if it is able to replace decaying components of its membrane. Depending on the kinetics of generation ($V_{gen}$) and decay ($V_{dec}$), this system will be in homeostasis ($V_{gen} = V_{dec}$), grow ($V_{gen} > V_{dec}$) or die ($V_{gen} < V_{dec}$) (Luisi 2003). Autopoietic systems depend only on open dissipative processes that allow the yielding of energy from the uptake and use of nutrients. They even do not require hereditary material, but it is easy to incorporate nucleic acids and enzymes into this theoretical framework (Luisi 2003).

From such simple forms of synthetic life to bacteria "rebooted" with a chemically synthesized genome (Gibson et al. 2010) is a large jump. While the former are close to non-living matter, the latter are indistinguishable from natural bacteria of the same species. This makes it necessary to devise a scheme, in which the different life forms that are produced by the diverse approaches of synthetic biology can be described and compared. This will also help to conceive risk assessment strategies and to give specific recommendations for cellular constructs generated in the diverse fields of synthetic biology. Here, we propose a layer model that should help to get a better picture of synthetic biology and to describe its different approaches.

## 2.3 A Layer Model to Describe and Classify Natural and Synthetic Organisms in Different Fields of Synthetic Biology

All approaches of synthetic biology outlined above are dealing with living or life-like systems thus expanding the number of potential designs to a great extent. Synthetic organisms will be created that have not been evolved and thus can be regarded as being beyond natural (Elowitz and Lim 2010). Here we develop a model, which will allow classifying the different artificial living systems according to their position within or outside of the realm of the natural life forms. All existing life forms on earth have arisen by natural evolution. Due to Darwinian evolution by random mutation and selection, natural "life" as it exists on earth is also highly contingent since many crucial inventions that enabled organic life on earth have been "invented" only once (for example the genetic code). It is commonly accepted that all living beings can be traced back to a "last universal common ancestor" (LUCA) and that all extant species are connected by a common genealogy (Kyrpides et al. 1999; Lazcano and Forterre 1999; Woese 1998).

However, from a global point of view, existing life forms represent only a very tiny fraction of all "potential" living beings including those that have not yet been evolved. All existing species depend on DNA as its hereditary material and use (with only very few exceptions) the same (universal) genetic code. Thus, every species is unambiguously defined by its genome sequence. Since DNA sequences consist of four different bases (A, C, G, T) and are of various lengths—typically they range from a few million base pairs to several billion—the number of potential genome sequences is almost infinite. If each (actual or potential) genome sequence is represented by a single point on a layer, all points of this layer represent the sequence space of all organisms that share a common genetic code. Within such a layer, the number of potential living organisms is infinite, since it equals the number of all possible genome sequences. Every point on the layer corresponds to a single species, either to one that exists in nature or to one that have not (or not yet) evolved. One has to point out, that within this layer of natural species exchange of genetic information is possible since all species share a common genetic system, the universal genetic code. Since the number of extant species is small if compared to the number of all possible genome sequences the area covered by natural is tiny and resembles more small islands in a vast empty space (see Fig. 2.1).

According to the modern evolution theory, species evolve by mutation, selection and genetic drift (Kimura 1983). Therefore all new species are closely related to their progenitors and contain only a few number of sequence deviations. Thus, the dynamics of evolutionary change results in a slow but continuous spreading of species that explore the neighbouring uncharted territory of the sequence space. Synthetic biology now provides the new and unique possibility to create any of potentially viable organisms within a layer while both natural evolution and gene technology depend more or less on the limited amount of present-day organisms and genomes.

Since the total number of potential organisms corresponds to the number of potential genomes, it thus equals the number of possible DNA sequences. Simple calculations demonstrate that in the previous billions of years only a very tiny fraction of all potential genomes has been realized by evolution. This is due to the fact that the combinatorial number of sequences is beyond our imagination. Genomes encode thousands of proteins but even for a rather short peptide of only 100 amino acids $20^{100}$ different sequences are possible. Therefore "the ratio between the possible (say $20^{100}$) and the actual chains [realized in nature by evolution] (say $10^{15}$) corresponds approximately to the ratio between … all the grains of sand in the vast Sahara and a single grain" (Luisi 2006). This does not imply

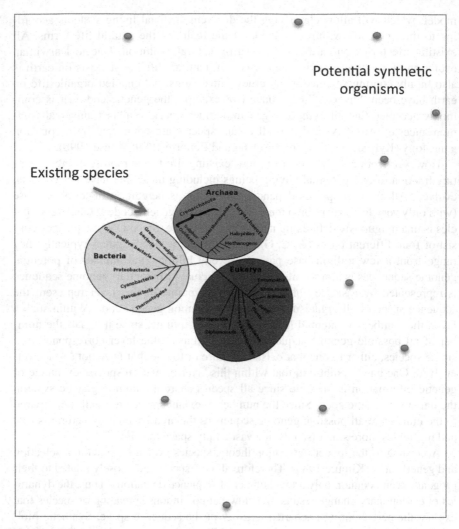

**Fig. 2.1** Although life on earth exists for more than 3 billion years, the number of genomes that have been realized by evolution corresponds only to a tiny fraction of the number of possible genomes. Synthetic biology is able to create synthetic organisms that are not derived from pre-existing organisms

that every point on the unlimited sequence space immediately corresponds to a putative living being. In contrary, one has to assume that the vast majority of all potential organisms are not able to "live" in the common sense. Since each point of the sequence space corresponds to one (random) sequence, the chance that it represents a successful living being, would equal that of a monkey typing randomly on a typewriter to reproduce the works of Shakespeare to quote a famous saying.

If the infinite sequence space of potential genomes is depicted as a two-dimensional layer, than only a tiny space is filled by the existing natural genomes that are all connected by evolutionary trajectories (Fig. 2.1). In contrast to natural evolution that explores the remaining space only in small mutational steps all starting from the existing species, synthetic biology can make immediate use of the complete sequence space, because it does neither depend on mutations nor on any previous gene sequences. The ability to read (sequence) and write (synthesize) DNA in large scale allows to design any sequence and even to assemble whole genomes. This has liberated synthetic biology from the direct connectivity to existing species (Fig. 2.1).

The layer model visualizes the large sequence space that now can be explored by designing novel genomes. In addition, the layer model is also able to accommodate attempts to create orthogonal life forms that are separated from the natural world by a "genetic firewall" (Marlière 2009). Those synthetic organisms that differ from natural ones in their biological and chemical nature can be placed on separate layers. Layers are defined such that within a certain layer all organisms use the same genetic material and use the same genetic code. This allows unlimited exchange of genetic material and information within a layer. One has to point out, that all biological systems created by the engineering branch of Synthetic Biology are based on DNA and the standard genetic code and therefore able to exchange genetic material not only between each other but also with the existing natural organisms. Between the layers no such exchange is possible because they differ in their chemical composition and/or in their genetic code. Therefore organisms that are placed on different layers are genetically isolated from each other and thus have to be regarded as orthogonal. The use of non-standard bases or reassigned

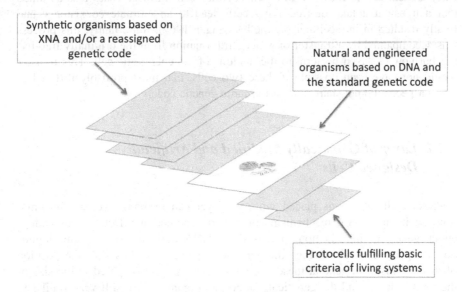

Synthetic organisms based on XNA and/or a reassigned genetic code

Natural and engineered organisms based on DNA and the standard genetic code

Protocells fulfilling basic criteria of living systems

**Fig. 2.2** The existing living organisms, produced by natural evolution, cover only a very tiny area of the nearly unlimited space of all potential organisms. The infinite sequence space of potential organisms sharing a common genetic code is indicated by a single layer. Beyond the natural form of life, characterized by DNA as genetic material and the standard genetic code, many other sequence spaces can be imagined, that differ in the genetic code, or in the chemical nature of the hereditary material. While exchange of genetic information is feasible within a plane, no such transfer is possible between different layers

genetic codes in orthogonal synthetic biology is visualized in our model as separate layers of possible genomes.

The layer model is not hierarchical and being placed in another layer does not indicate that some organisms are higher developed than others. However, there might be some layers with a more restricted sequence space, e.g. due to a reduced genetic code encoding a smaller number of amino acids. Such layers most probably correspond to early stages of natural evolution but would also apply to artificial protocells with a reduced genomic content (Loakes and Holliger 2009) (Fig. 2.2).

According to this model, a single layer of the sequence space contains only species that share a common molecular basis of metabolism and reproduction. In particular, all natural organisms belong to one single layer, which might be termed the "natural" layer. It is defined by the use of the "universal" genetic code, which obviously has evolved only once.[1] The universality of the natural genetic code has an important consequence: since it allows the meaningful exchange of genetic information not only between closely related species, but also between those that are only distantly related and may share a common ancestor dating back millions of years. Such events, termed horizontal gene transfer, may occur in nature only infrequently. Nevertheless, they have shaped many species and are an important factor in evolution. Furthermore, several times during evolution novel species have been generated by endosymbiosis events in which complete cells were taken up by another species. The most spectacular case is the fusion of a bacterium with an early archeal cell to form the first "eukaryotic" cell. In these cases it is assumed that a transient uptake of free-living cells lead to an intimate coexistence and finally resulted in intracellular organelles (Sagan 1967). Importantly, this process was accompanied by a transfer of genetic information from the originally free-living but now enslaved bacteria to the nucleus of the eukaryotic cell. This transfer would not have been possible if these two cells, that most probably differed in many aspects, did not share at least the same genetic code.

## 2.3.1 Layer of Genetically Modified and Artificially Designed Cells

Therefore, all organisms placed within a layer can exchange genetic information, be it via natural mechanisms or genetic engineering. Deletion of endogenous genes, or introduction of additional DNA derived from a heterologous donor organism, would change the genome sequence and thus shift the position of this organism. In general these alterations are small if compared to the size of the total genome, and the genetically modified organisms are still very similar to their natural progenitors. But with increasing number of transferred genes and size of transformed DNA the relationship with the original organism dwindles. If a

---

[1]Interestingly there exist a few exceptions e.g. in Candida yeasts (Ohama et al. 1993; Santos and Tuite 1995), but these species use a code with some small deviations, i.e. the code is not totally unrelated.

large number of genes, originating from many different species, is used to create a "chimeric" new organism containing an artificially designed genome, this genome sequence has to be placed within the layer of natural species but somewhere in the open space as yet uncharted by evolution. Although the biological and biochemical basis of such synthetic creatures does not differ from their natural counterparts, one may regard such living cells as "Life beyond natural Life" (Elowitz and Lim 2010). Within our scheme it becomes obvious, that there is no principal difference between organisms created by natural evolution, natural mechanisms of horizontal gene transfer, gene technology or by engineering synthetic biology: all are based on DNA as genetic material and the universal genetic code. But the groundbreaking "creation of a bacterial cell controlled by a chemically synthesized genome" has demonstrated: Current synthetic biology does not any longer depend on natural sources of DNA to generate a living organism. It was only the sequence information that was used to direct chemical synthesis and assembly of a complete bacterial chromosome (Gibson et al. 2010). However, this implies that any sequence designed in the computer can be realized by chemical synthesis and may serve as a blueprint for a novel bacterial species. There are no principal technical barriers, at least on the level of genome synthesis, preventing us from designing a synthetic organism whose genome would correspond to any point on the layer, independently whether this designed organism would be viable or not.

Thus, it is not only that all existing biological species can be used as a starting point for genetic modification but also that the vast space of not yet evolved organisms can be explored by synthetic biology. Many critics of the Venter experiments have complained that a mere exact synthetic copy of a natural bacterium just confirms what has long been known, that it is the DNA sequence, which determines the identity and the phenotype of a species. However, the aim of that study was less to demonstrate the creation of novel forms of life, but more to show that it is feasible to synthesize any given genome sequence and to bring it to life with the help of a DNA-free cell. Admittedly, there are and will be many technical obstacles to use this technique for implementing synthetic genomes that are not closely related to existing ones. One of the biggest challenges will be which cells can be used to accept a chemically synthesized genome. It has to be compatible with the incoming DNA to allow for the replacement of endogenous ribosomes by the novel ones. Craig Venter calls this step the "rebooting" of a cell with a new genome. However, "rebooting" may require a high level of genomic compatibility, setting tight constraints on the fantasy of genome designers.

The layer model demonstrates that there are no explicit distinctions between organisms that have been shaped by natural processes such as mutation, recombination and horizontal gene transfer, and those produced by genetic modification or even by introducing a chemically synthesized genome. Even if cells were generated by extreme genetic engineering using many different sources of natural donor DNA or by synthetic biology on the basis of a complete new genome design, in all cases these organism share the same principles of metabolism and inheritance. Therefore they can interact with each other not only in the environment but also by transfer and exchange of genetic information. This has been interpreted by many to state that synthetic biology essentially is nothing but gene technology. Taking an extreme position, Rob Carlson even claims that natural processes occurring during evolution can be regarded as "technology", as stated in his book "Biology is Technology" (Carlson 2009). He describes "life" as we see it today as organisms that exploit each other to survive and make energy in the Earth's newly formed atmosphere. Eukaryotic organisms utilize mitochondria and chloroplasts acquired through symbiosis for their own purpose. In this sense, bacteriophages appear to be perfect synthetic biologists. Injection of a short piece of viral DNA is sufficient to reprogram a bacterial cell to produce several hundred bacteriophage copies within less than an hour.

Nevertheless, there are good reasons to stick to the traditional scheme of distinguishing organisms derived by natural modes of gene transfer from those modified by classic gene technology or constructed as artificially designed synthetic cells. Especially in the discussion of potential risks for health or environment, we depend on experience and knowledge of both the probability and the damage that might be caused by novel species. We know that natural organisms have existed for as long as life has existed on earth. This does not necessarily imply that these organisms are by any means harmless; mankind was and is always confronted with emerging diseases, including pandemic ones. But we know that all existing organisms, including ourselves, have been shaped by this long record of historical challenges, and we are confident that our immune system will be able to cope with these challenges as it has developed during evolution just for this purpose (if one may use such teleological terminology as a surrogate for selection and fitness). For organisms genetically modified by classic gene technology, it is the close similarity and relationship to existing biological species that can be used to assess potential risks and to decide at which safety level experiments have to be performed. These criteria are not available for synthetic cells that are either composed of DNA fragments originating from many different natural sources, or that have been rationally designed with the help of computers (see Chap. 3 on risk).

Engineering biology aims at a simplification of biological systems into modular and abstract networks to facilitate re-engineering of living systems and even make the technical "synthesis of life" feasible. Philosophically speaking, the belief in an "artificial creation of life" is based on the rigid determinism and reductionism of 19th century materialistic western philosophy. This is best illustrated by a lecture entitled "Life", held in Hamburg, Germany, in 1911 where the American physiologist Jacques Loeb proclaimed his reductionist view that sees living organisms as chemical machines (see Chap. 4). There is indeed a long history of ingenious automatic devices already in the late classical period and during medieval times, not only for popular entertainment but also to serve religious purposes and which later influenced the mechanical natural philosophy of the Enlightenment. For example, Descartes maintained that animals (and the human body) are "automata", mechanical devices differing from artificial devices only in their degree of complexity. With a similar belief in mind, Loeb proclaimed the goal of a controlled and useful design of "synthetic life" by forming new combinations of the elements of living nature by engineering. The engineering version of Synthetic Biology is also reflected in the prescient imagination of the French biologist Stephane Leduc, who proclaimed a future quantitative biology, which—in analogy to theoretical physics—should give a precise description of organisms whose behaviour is not only completely quantifiable but also predictable and controllable.

From the engineering perspective of modern synthetic biology, a cell is regarded as a complex system of devices and functions that can be reconstructed by rational design. A major goal is the reduction of the complexity of natural cells, which appears to be an unnecessary complication due to historic contingencies. The commonly accepted generalized strategy for this approach is the design and construction of a standardized minimal cell. Such cells can then be used as a chassis to set up artificially designed cells that fulfil useful functions that are novel and have not been seen in nature. The minimal cell itself is equipped only with the basic life-sustaining functions but has been stripped of everything which appears in the eyes of the engineers as contingent or unnecessary. Although there exist no bona fide minimal cells yet, a number of different approaches have been discussed (Acevedo-Rocha et al. 2013). Popular approaches to exploring the "minimal requirements for cellular life" include comparative genomics and massive genome mutagenesis (Forster and Church 2006). Synthetic biologists usually approach this problem by considering "minimal genetic requirements" or by employing information theory to living systems. But, one has also to think of the chemical invariance of living beings: a finite number of basic chemical building blocks (metabolites, amino acids, and nucleotides) participate in endless cycles of transformation between plants, animals and microorganisms in the biosphere. At least one can imagine that one day such a minimal cell could gain the status of a model system like *E. coli*. If one gains enough experience working with such an established minimal cell, it may even be less risky than working with natural cells. This concept could pave the way to a standardized synthetic biology.

## 2.3.2 Layer of Orthogonal Cells (Xenobiology)

One of the basic assumptions of orthogonality and xenobiology is that the molecular nature of the extant living organism is only one solution to the problem of creating life. If evolution were rerun on our planet, most probably another sequence of molecular events would have occurred. Therefore the specific chemistry of living beings is for a large part due to historical contingencies. In the very moment where the first successful life forms emerged on earth, it was no longer possible to switch to other molecules of life. This suggests, however, that at least potentially there should exist many more molecular structures able to constitute living matter. This view was supported by some early and recent experiments where natural amino acids and co-factors were replaced by related substances without severe loss of function (Cowie and Cohen 1957; Lemeignan et al. 1993; Ma et al. 2014). Therefore the space of yet unexplored alternative life forms is enormous.

This raises the question whether these life forms can indeed be regarded as "alternative" or "beyond natural" (Elowitz and Lim 2010). Here one has to consider that xenobiotic life forms are more or less explicitly modelled on existing living beings. For example genetic information is stored by XNA, which resembles natural DNA not only in its base pairing principle but also in its double helical structure. But this chemical mimicking of natural life, in combination with the inherent genetic firewall of xenobiotic organisms, provides at least one significant advantage to engineering synthetic biology. With regard to their biological safety orthogonal organisms are easier to handle since by design they are unable to exchange any genetic information with natural organisms. They are for example resistant to phages, which could become an advantage in production. It is argued that the farther they are away from natural organisms the safer they are (Marlière 2009; Budisa 2014). This is in sharp contrast to the engineering branch of synthetic biology that argues that artificial life forms constructed on the basis of minimal cells and standardized modules are safer because their behaviour will be much better predictable. Therefore, xenobiotic cells can keep some of the vitality and unpredictability of their natural counterparts.

In contrast to minimal cells that are at least derived from normal cells, it is not so easy to decide whether a cell is still alive if many or all important biological functions are exerted by similar but not identical chemical compounds. If one sticks to a definition of life, which also refers to its material basis, for example that life is defined as being based on DNA as genetic material, then xenobiotic life forms using alternative XNA for information storage cannot be regarded as alive. However, many biological definitions do not make any material constraints but refer only to functional attributions (cf. Chap. 4 by Toepfer). Therefore, replacing certain (or all) parts by those of different chemical composition would not affect the ability of the system to be alive. Even from a practical aspect this is evident, since all realistic approaches to the construction of xenobiotic cells conceptually start with the chemical transformation of an existing organism. Into this undoubtedly living entity, novel components are inserted which then take over certain cellular functions (see Schmidt and de Lorenzo 2012). However, if DNA is replaced by XNA, such an organism would no longer fit into the layer of natural species. Since exchange of genetic information is no longer possible due to the different chemical nature of its material, organisms would belong to a parallel layer, which is neither intersecting nor overlapping with the natural layer.

Although initial steps have been achieved (Malyshev et al. 2014) organisms containing XNA instead of DNA are still far from being realized. Two types of orthogonal organisms, however, might be much easier to achieve: recoded cells with a reassigned genetic code and mirror-like cells based on stereoisomers of natural compounds. In principle, reassignment of only a few codons is sufficient to exclude any meaningful exchange of genetic information. There is good evidence that large-scale reassignment of the genetic code will not be easy to achieve. This is due to the structural constraints in enzymatic realization of different codes, for example in the biochemical structure of proteins involved in charging tRNAs with amino acids. Interestingly, some naturally occurring species (e.g. *Candida albicans*) have evolved reassignment of a single codon demonstrating that genetic recoding of cells is feasible (Ohama et al. 1993; Santos and Tuite 1995)

An interesting alternative to the difficult tasks of xenobiology or genetic recoding is the construction of mirror-like cells (Church and Regis 2012). Such cells are predicted to behave identically to natural ones since they are based on mirror-like versions of natural compounds. Most organic molecules used in biological systems display chirality, and in all cases nature uses only one of the two possible stereoisomers. Since it is known from physics and chemistry that all properties and reactions are independent of chirality, one can predict that a mirror-like cell will display exactly the same properties and phenotypes as its natural counterpart. However, such cells would fully depend on left-handed compounds whereas a normal cell uses right-handed ones and vice versa. For biotechnology this might be even an advantage. Since all existing natural species are subject to parasite attacks, being mirror-like makes a bacterial cell invulnerable to the attack of bacteriophages. George Church even envisions the creation of mirror-like humans who would then constitute a "parallel" humankind (Church and Regis 2012). With regard to our layer model there exists exactly one orthogonal layer of mirror-like organisms; for every single existing species there exists exactly one that is of opposite chirality. In contrast, with regard to codon reassignments there are almost infinite orthogonal layers if one considers all possible reassignments of the genetic codes. One can even imagine that companies construct their "individual" (and maybe patentable) cells, which are characterized by their highly specific genetic code. These cells are "labelled" by their genetic code and can always be traced back to their inventors in case of alleged misuse of intellectual property.

Although in principle every species might be equipped with an alternative genetic code, there might be serious constraints: regulatory elements such as enhancers depend on DNA sequence. If located in coding regions the sequence of such elements will be altered upon recoding. Therefore it is not guaranteed that replacement of many or even a few codons is compatible with functionality; fitness could be severely decreased. But genetic recoding might be interesting in the construction of minimal cells, which are reduced in their biological complexity anyway. One can imagine different sets of minimal cells constructed on the basis of alternative genetic codes. These constraints are much greater upon the introduction of XNA. Here it is already extremely difficult to establish such an organism. Current scenarios imply a transition from natural to xeno-organisms by gradually replacing natural DNA with its chemical counterpart XNA (Schmidt 2010). This resembles the metaphor of the Delphic boat (Danchin 2003, 2009), where it is not clear whether it is the same after all the planks have been replaced during its long journey. In xenobiology it would be the property of being alive which is maintained (or not), even if all the chemical components are replaced by different versions.

## 2.3.3 Layer of Simple Life Forms and Protocells

Protocells are cell-like structures that are spatially delimited by a membrane boundary and contain biological material that can be replicated. Ideally they consist of a self-assembling chemical system capable of reproduction. Current research in this field is mainly concerned with the fundamental questions of the origin of life (Rasmussen et al. 2009). The crucial goal is to understand and experimentally master the transition from complex abiotic chemistry to simple biology that is to enable the emergence of complex chemical assemblies capable of fulfilling the criteria for life. In other words, to perform the transition from "non-life" to "life" one needs to define at least basic terminology: "non-life" usually means the maintenance of chemical rules (chemical equilibrium and reaction rates) whereas "life" is described as a system that maintains biological rules (selective pressures from the environment and replication rates).

Another motivation for protocell research is the "artificial life" (AL) field which traditionally includes *in silico* design of systems constructions that exhibit lifelike behaviour ("Soft" AL) and even robotics ("Hard" AL). In this context, "fluid" (or "wet") AL involves the creation of lifelike protocells, based on aqueous carbon chemistry in water with repeatable sets of autocatalytic chemical cycles that are properly coordinated. These autocatalytic cycles include replication of informational material coupled with internal metabolic reactions in a self-maintained manner within the membrane boundary. In an ideal case, the conditions for the coordination of the growth and shape changes of the cell's membrane (that is cell growth and division) are defined. At the level of simple physicochemical laws, the processes of vesicle growth and reproduction occur as a consequence of breaking the spatial symmetry of a synthetic protocell. Finally, protocell research can be employed for the design of drug delivery systems in medicine, for example the encapsulation of drug cocktails in liposomes and various nanoparticle.

If one questions whether protocells are alive or not, the answer very much depends on the criteria that are used to distinguish between living and non-living systems (see Toepfer, Chap. 4). Traditionally, living beings have to fulfil a considerable number of criteria derived from the description of existing biological organisms. In protocell research it is often a single unifying aspect of living systems, which is used to distinguish living systems from non-living ones. Autopoiesis has been proposed to serve as such a unifying concept (Luisi 2006). Autopoiesis is the ability to maintain a system from within by replacing components that are degraded by chemical processes. Such an autopoietic process results in stable maintenance if the number of replaced components equals the number of degraded ones. If the system produces more components than are degraded, the system will grow. This definition does not depend on reproduction or genetic information but resembles more the chemoton definition of life (Gánti 1975) which is modelled along dynamic chemical equilibria. Protocells have a dualistic nature; on the one hand they can be described as a complex chemical system but on the other hand they can be seen as simple forms of life (Rasmussen et al. 2009).

In the layer model, the size of a layer is determined by the number of possible genome sequences. Typically, synthetic protocells carry no heritable material at all (purely metabolic networks) or only very short polymers capable of self-replication (Mansy and Szostak 2009). Therefore, protocells are usually unable to exchange any kind of "genetic" information with each other. This indicates that in our layer model every protocell represents its own small island, separated from other protocells that operate by different metabolic mechanisms. Layers, in a strict sense, are thus only present if some kind of hereditary system will be established allowing the exchange of information between different protocells. Admittedly, with such simple life forms the layer model reaches its limits. But this may actually reflect the unique position that protocells occupy if compared both to natural and engineered synthetic cells.

## 2.4 Conclusions

One of the major goals of synthetic biology is the design of novel living systems. These can be based on defined minimal cells that are derived from naturally existing cells but with reduced complexity. In essence, this strategy follows traditional engineering principles: reduction of complexity by modularization in combination with quantitative estimation of module parameters is expected to allow total control of such highly integrated "systems". Whether the behaviour of such synthetic cells indeed becomes more predictable remains to be seen since also purely technical systems turned out to be less controllable as originally planned especially if they surpass a certain level of complexity.

The generation of orthogonal living systems follows an alternative rationale: Here it is the genetic incompatibility of these organisms with natural species that confers a certain level of safety. But this is paid by the significantly higher level of unfamiliarity that characterizes such xenobiological organisms. Therefore, in the moment, it is difficult to judge whether this strategy will ever provide a practicable road to synthetic cells that can be used also outside the laboratories.

Thirdly, protocells might be an interesting alternative for certain simple applications that do not require fully living cells. Here it is the defined chemical composition and the inability of these systems to compete with complex organisms that provides some opportunities both with regard to safety considerations and applications.

## References

Acevedo-Rocha CG, Fang G, Schmidt M, Ussery DW, Danchin A (2013) From essential to persistent genes: a functional approach to constructing synthetic life. Trends Genet 29:273–279
An W, Chin JW (2009) Synthesis of orthogonal transcription-translation networks. Proc Natl Acad Sci USA 106:8477–8482

Andrianantoandro E, Basu S, Karig DK, Weiss R (2006) Synthetic biology: new engineering rules for an emerging discipline. Mol Syst Biol 2. doi:10.1038/msb4100073

Attwater J, Holliger P (2014) A synthetic approach to abiogenesis. Nat Meth 11:495–498

Benner SA, Sismour AM (2005) Synthetic biology. Nat Rev Genet 6:533–543

Benner SA, Chen F, Yang ZY (2011) Synthetic biology, tinkering biology, and artificial biology: a perspective from chemistry. In: Luisi LP, Chiarabelli C (eds) Chemical synthetic biology. Wiley, Chichester, pp 69–106

Bohannon J (2011) The life hacker. Science 333:1236–1237

Budisa N (2014) Xenobiology, new-to-nature synthetic cells and genetic firewall. Curr Org Chem 18:936–943

Cameron DE, Bashor CJ, Collins JJ (2014) A brief history of synthetic biology. Nat Rev Microbiol 12:381–390

Campos L (2009) That was the synthetic biology that was. In: Schmidt M, Keller A, Ganguli-Mitra A, de Vriend H (eds) Synthetic biology. The technoscience and its social consequences. Springer, Heidelberg, pp 6–22

Canton B, Labno A, Endy D (2008) Refinement and standardization of synthetic biological parts and devices. Nat Biotechnol 26:787–793

Carlson R (2009) Biology is technology. The promise, peril, and new business of engineering life. Harvard University Press, Cambridge, MA

Chan LY, Kosuri S, Endy D (2005) Refactoring bacteriophage T7. Mol Syst Biol 1(2005):0018. doi:10.1038/msb4100025

Chin JW, Cropp TA, Anderson JC, Mukherji M, Zhang Z, Schultz PG (2003) An expanded eukaryotic genetic code. Science 301:964–967

Church GM, Regis E (2012) REGENESIS: how synthetic biology will reinvent nature and ourselves. Basic Books, New York

Cohen SN, Chang AC, Hsu L (1972) Non-chromosomal antibiotic resistance in bacteria: genetic transformation of Escherichia coli by R-factor DNA. Proc Natl Acad Sci USA 69:2110–2114

Cohen SN, Chang ACY, Boyer HW, Helling RB (1973) Construction of biologically functional bacterial plasmids in vitro. Proc Natl Acad Sci USA 70:3240–3244

Cowie DB, Cohen GN (1957) Biosynthesis by Escherichia coli of active altered proteins containing selenium instead of sulfur. Biochim Biophys Acta 26:252–261

Danchin A (2003) The Delphic boat. What genomes tell us. Harvard University Press, Cambridge

Danchin A (2009) Information of the chassis and information of the program in synthetic cells. Syst Synth Biol 3:125–134

Deplazes A, Huppenbauer M (2009) Synthetic organisms and living machines. Syst Synth Biol 3:55–63

Eigen M, Schuster P (1977) The hypercycle: a principle of natural self-organization. Part A: emergence of the hypercycle. Naturwissenschaften 11:541–565

Elowitz M, Lim WA (2010) Building life to understand it. Nature 486:889–890

Endy D (2005) Foundations for engineering biology. Nature 438:449–453

ETC Group (2007) Extreme genetic engineering: an introduction to synthetic biology. ETC Group

Forster AC, Church GM (2006) Towards synthesis of a minimal cell. Mol Syst Biol. 2:45

Gánti T (1975) Organization of chemical reactions into dividing and metabolizing units: the chemotons. Biosystems 7:15–21

Gibson DG, Glass JI, Lartigue C, Noskov VN, Chuang RY, Algire MA, Benders GA, Montague MG, Ma L, Moodie MM, Merryman C, Vashee S, Krishnakumar R, Assad-Garcia N, Andrews-Pfannkoch C, Denisova EA, Young L, Qi ZQ, Segall-Shapiro TH, Calvey CH, Parmar PP, Hutchison CA 3rd, Smith HO, Venter JC (2010) Creation of a bacterial cell controlled by a chemically synthesized genome. Science 329:39–50

Hammerling MJ, Ellefson JW, Boutz DR, Marcotte EM, Ellington AD, Barrick JE (2014) Bacteriophages use an expanded genetic code on evolutionary paths to higher fitness. Nat Chem Biol 10:178–180

Hirao I, Kimoto M (2012) Unnatural base pair systems toward the expansion of the genetic alphabet in the central dogma. Proc Jpn Acad Ser B Phys Biol Sci. 88:345–367

Hoesl MG, Budisa N (2012) Recent advances in genetic code engineering in *Escherichia coli*. Curr Opin Biotechnol 23:751–757

Hunter P (2013) XNA marks the spot. EMBO Rep 14:410–413

Isaacs FJ, Carr PA, Wang HH, Lajoie MJ, Sterling B, Kraal L, Tolonen AC, Gianoulis TA, Goodman DB, Reppas NB, Emig CJ, Bang D, Hwang SJ, Jewett MC, Jacobson JM, Church GM (2011) Precise manipulation of chromosomes in vivo enables genome-wide codon replacement. Science 333:348–353

Jackson DA, Symons RH, Berg P (1972) Biochemical method for inserting new genetic Information into DNA of simian virus 40: Circular SV40 DNA molecules containing lambda phage genes and the galactose operon of *Escherichia coli*. Proc Natl Acad Sci USA 69:2904–2909

Joyce GF (2002) The antiquity of RNA-based evolution. Nature 418:214–221

Juhas M, Eberl L, Church GM (2012) Essential genes as antimicrobial targets and cornerstones of synthetic biology. Trends Biotechnol 30:601–607

Kimura M (1983) The Neutral Theory of Molecular Evolution. Cambridge University Press, Cambridge

Knight T (2003) Idempotent vector design for standard assembly of biobricks. http://dspace.mit.edu/handle/1721.1/21168 Accessed 08 Mar 2015

Kyrpides N, Overbeek R, Ouzounis C (1999) Universal protein families and the functional content of the last universal common ancestor. J Mol Evol 49:413–423

Lajoie MJ, Rovner AJ, Goodman DB, Aerni HR, Haimovich AD, Kuznetsov G, Mercer JA, Wang HH, Carr PA, Mosberg JA, Rohland N, Schultz PG, Jacobson JM, Rinehart J, Church GM, Isaacs FJ (2013) Genomically recoded organisms expand biological functions. Science 342:357–360

Lazcano A, Forterre P (1999) The molecular search for the last common ancestor. J Mol Evol 49:411–412

Leduc S (1912) La biologie synthétique. In: Poinat A (eds) Études de biophysique. Paris

Lemeignan B, Sonigo P, Marlière P (1993) Phenotypic suppression by incorporation of an alien amino acid. J Mol Biol 231:161–166

Loakes D, Holliger P (2009) Darwinian chemistry: towards the synthesis of a simple cell. Mol BioSyst 5:686–694

Liu CC, Schultz PG (2010) Adding new chemistries to the genetic code. Annu Rev Biochem 79:413–444

Luisi PL (1998) About various definitions of life. Orig Life Evol Biosphere 28(613–622):1998

Luisi PL (2003) Autopoiesis: a review and a reappraisal. Naturwissenschaften 90:49–59

Luisi PL (2006) The emergence of life: from chemical origins to synthetic biology. Cambridge University Press, Cambridge

Luisi PL (2007) Chemical aspects of synthetic biology. Chem Biodivers 4:603–621

Ma Y, Biava H, Contestabile R, Budisa N, di Salvo ML (2014) Coupling bioorthogonal chemistries with artificial metabolism: intracellular biosynthesis of azidohomoalanine and its incorporation into recombinant proteins. Molecules 19:1004–1022

Malyshev DA, Dhami K, Lavergne T, Chen T, Dai N, Foster JM, Correa IR Jr, Romesberg FE (2014) A semi-synthetic organism with an expanded genetic alphabet. Nature 509:385–388

Mandell DJ, Lajoie MJ, Mee MT, Takeuchi R, Kuznetsov G, Norville JE, Gregg CJ, Stoddard BL, Church GM (2015) Biocontainment of genetically modified organisms by synthetic protein design. Nature 518:55–60

Mansy SS, Szostak JW (2009) Reconstructing the emergence of cellular life through the synthesis of model protocells. Cold Spring Harb Symp Quant Biol 74:47–54

Marlière P (2009) The farther, the safer: a manifesto for securely navigating synthetic species away from the old living world. Syst Synth Biol 3:77–84

Miller SJ (1953) A production of amino acids under possible primitive earth conditions. Science 117:528–529

Ohama T, Suzuki T, Mori M, Osawa S, Ueda T, Watanabe K, Nakase T (1993) Non-universal decoding of the leucine codon CUG in several Candida species. Nucleic Acids Res 21:4039–4045

Pereto J (2012) Out of fuzzy chemistry: from prebiotic chemistry to metabolic networks. Chem Soc Rev 41:5394–5403

Pinheiro VB, Taylor AI, Cozens C, Abramov M, Renders M, Zhang S, Chaput JC, Wengel J, Peak-Chew SY, McLaughlin SH, Herdewijn P, Holliger P (2012) Synthetic genetic polymers capable of heredity and evolution. Science 336:341–344

Rasmussen S, Bedau MA, Chen L, Deamer D, Krakauer DC, Packard NH, Stadler PF (eds) (2009) Protocells: bridging nonliving and living matter. MIT Press, Cambridge

Reardon S (2011) Visions of synthetic biology. Science 333:1242–1243

Sagan L (1967) On the origin of mitosing cells. J Theor Biol 14:255–274

Santos MA, Tuite MF (1995) The CUG codon is decoded in vivo as serine and not leucine in *Candida albicans*. Nucleic Acids Res 23:1481–1486

Schmidt M (2010) Xenobiology: a new form of life as the ultimate biosafety tool. BioEssays 32:322–331

Schmidt M, de Lorenzo V (2012) Synthetic constructs in/for the environment: managing the interplay between natural and engineered biology. FEBS Lett 586:2199–2206

Schwille P (2011) Bottom-up synthetic biology: engineering in a tinkerer's world. Science 333:1252–1254

Stano P, Luisi PL (2013) Semi-synthetic minimal cells: origin and recent developments. Curr Opin Biotechnol 24:633–638

Szybalski W, Skalka A (1978) Nobel prizes and restriction enzymes. Gene 4:181–182

Trafton, A (2011) Rewiring cells: how a handful of MIT electrical engineers pioneered synthetic biology. MIT Technology Review. http://www.technologyreview.com/article/423703/rewiring-cells. Accessed 9 Mar 2015

Voigt C (2011) Access through refactoring: Rebuilding complex functions from the ground up. Presentation given at the March 14-15, 2011, public workshop, "Synthetic and Systems Biology," Forum on Microbial Threats, Institute of Medicine, Washington, DC

Wang HH, Isaacs FJ, Carr PA, Sun ZZ, Xu G, Forest CR, Church GM (2009) Programming cells by multiplex genome engineering and accelerated evolution. Nature 460:894–898

Wang K, Sachdeva A, Cox DJ, Wilf NW, Lang K, Wallace S, Mehl RA, Chin JW (2014) Optimized orthogonal translation of unnatural amino acids enables spontaneous protein double-labelling and FRET. Nat Chem 6:393–403

Woese CR (1998) The universal ancestor. Proc Natl Acad Sci USA 95:6854–6859

Yeh BJ, Lim WA (2007) Synthetic biology: lessons from the history of synthetic organic chemistry. Nat Chem Biol 3:521–525

# Chapter 3
# Old and New Risks in Synthetic Biology: Topics and Tools for Discussion

Margret Engelhard, Michael Bölker and Nediljko Budisa

**Abstract** When approaching the discussion on possible risks that might be elicited by synthetic biology we face two problems: The first one is the fluid definition of synthetic biology and the associated difficulties structuring the discussion. In the context of benefit/risk assessment this problem is even aggravated, since the label of synthetic biology is used ambiguously depending on political context. One possible tool to circumvent this problem is to reflect the realities of the research field and to focus on new features of synthetic biology that are relevant to the risk discussion. These new features are in particular the growing depth of intervention in the organism and, the decreasing familiarity of synthetic organisms that together with the high speed of technological development challenge established risk assessment systems. This leads to the second problem in the risk assessment: the increasing level of uncertainty associated with synthetic organisms that is due to our lack of knowledge of their behaviour in the environment which cannot be reduced by research within a relevant time of action. Thus, lacks of knowledge in combination with the transformative potential of synthetic biology are the main challenges ahead.

Emerging new technologies quickly provoke a discussion on potential new risks. In the case of synthetic biology—often considered as an extreme form of conflict-laden genetic engineering—front lines along the risk evaluation are rapidly

M. Engelhard (✉)
EA European Academy of Technology and Innovation Assessment GmbH,
Wilhelmstr. 56, 53474 Bad Neuenahr-Ahrweiler, Germany
e-mail: engelhar@hs-mittweida.de

M. Bölker
Philipps-Universität Marburg, Fachbereich Biologie, Karl-Von-Frisch-Str. 8,
35032 Marburg, Germany
e-mail: boelker@uni-marburg.de

N. Budisa
Department of Chemistry, TU Berlin, Müller-Breslau-Straße 10, 10623 Berlin, Germany
e-mail: nediljko.budisa@tu-berlin.de

© Springer International Publishing Switzerland 2016
M. Engelhard (ed.), *Synthetic Biology Analysed*, Ethics of Science
and Technology Assessment 44, DOI 10.1007/978-3-319-25145-5_3

51

developing. A focal point in this context is the judgment of the novelty and the framing of the technology. Are we dealing with new risks or old ones? Are risk assessment and management systems currently in place, sufficient to deal with synthetic biology or do we need to adapt them?

Many of these questions relate to the very nature of synthetic biology, which is discussed in this volume: whether it is a novel technology or just an extension of the well-known genetic engineering[1] and what are its perimeters. Thus, we have to consider these topics and analyze them in the light of risk evaluation. While synthetic biologists cope very well without a concise definition of their new science, its precise or imprecise scope can have huge effects if potential chances and risks of synthetic biology have to be assessed. E.g. depending on a narrower or wider definition some groups might be included or excluded from funding programs or might or might not fall under certain legal regimes.

This societal and political level might be one reason for the ambiguous labels and descriptions of synthetic biology in the risk discussion. Another significant contributing factor, which shapes the complex discussion, is the large diversity of disciplinary backgrounds of the scientists (Sect. 3.1.1). In general, active researchers feed their specific horizon of experience into the discussion and thus they have very different views on the scope, novelty and risk evaluation of synthetic biology.

We will argue that due to the ambiguity of the label "synthetic biology" (Sect. 3.1) it could be a helpful strategy to focus on the characteristic features of this new technology that might be relevant for evaluation (Sect. 3.3). Therefore, identification of specific features of synthetic biology and its subfields (e.g. the imminent loss of familiarity with synthetic organisms) that are relevant to risk evaluation might be a good way out of this dilemma. A second argument for compiling such a list of new features irrespective of a more rigid or ambiguous definition of synthetic biology is that this strategy remains flexible and can easily be adapted to new scientific developments. Nevertheless, the mosaic structure of synthetic biology makes it necessary to take the diversity of its subdisciplines into account for risk assessment.

## 3.1  Synthetic Biology in the Risk Discussion

One of the first obstacles in assessing the risks of synthetic biology lays in the reality of this scientific field itself. Its highly fractionated and mosaic-like character (Sect. 3.1.1) gives space to utilize the label synthetic biology in a very broad and ambiguous way: on the level of scope, continuity, novelty and time point of development.

---

[1]See also Chap. 2.

### 3.1.1  Background: The Mosaic-like Structure of the Field

The synthetic biology research field is multi-layered and diverse. Beyond the subdivisions of engineering, orthogonal and protocell branches, the plurality of scientific backgrounds of the protagonists further contributes to the structuring of the field. The synthetic biology research community consists of scientists trained in a vast variety of disciplines such as engineering, organic chemistry, systems biology, theoretical physics or computational science, to mention but a few. All of these fields provide their own concepts, metaphors, tools, and models, which are typically utilized by synthetic biologists if drawing analogies between the old and new field of inquiry (Knuuttila and Loettgers 2014). In addition, we can show[2] that research agendas of the individual scientist are connected to and driven by the individual disciplinary roots (Engelhard and Hagen 2012). John Glass of the Craig Venter Institute, for example, is a virologist by training and worked for a long time in polio and vaccinia research, before he moved into bacterial pathogenesis and genomics. He became interested in organisms called mycoplasmas, which are the smallest cells known to be capable of independent growth. He was caught up by the idea of defining a minimal cell, not knowing that this ever would be called synthetic biology. Later he became head of the synthetic biology group at the Craig Venter Institute and has since then focussed on the design and construction of synthetic viral and bacterial genomes. One important track of that research was the development of new techniques called genome assembly and genome transplantation that culminated in the first bacterial cell controlled by a chemically synthesized genome. This cell was called "synthetic" by the authors, because its genome was chemically synthesized rather than replicated from an existing template. This experiment is seen as an important step in the development of a system that would allow the construction of a minimal bacterial cell, which could lead to a better understanding of the first principles of cellular life (Gibson et al. 2010; Glass 2012) but also serve as a useful tool for biotechnological applications. Two recent publications by John Glass on synthetic vaccine virus generation (Dormitzer et al. 2013) and bacterial genome reduction (Suzuki et al. 2015) illustrate that disciplinary background persists for long time in the individual research agendas of synthetic biologists. This is only one example demonstrating the close connection between the scientific and academic training of the leading scientists and their current research agendas and also their visions of the future. This phenomenon stands for a common pattern in synthetic biology and some more examples are given in Table 3.1.

When studying the examples in Table 3.1, it becomes obvious that synthetic biology is characterized by a scientific pluralism that goes far beyond the

---

[2]Most of the provided information is based on a non-public expertise conducted by Margret Engelhard and Kristin Hagen for the German Parliament. The expertise (duration: 2/2012– 12/2012) is based on qualitative interviews with leading scientists from within synthetic biology, scientists that research on synthetic biology and active artists. Main content of the interviews were the current status of synthetic biology, its framing (also in comparison to genetic engineering), on xenobiology and protocell research, the individual research agendas, the role of DIY-biology and questions on potential risks.

**Table 3.1** Influence of the disciplinary background on the focus of the research agendas of 12 synthetic biologists (Engelhard and Hagen 2012)

| Disciplinary background | Focus of research in synthetic biology |
|---|---|
| Biochemistry, genetics | Reduction of signal complexity, signal-transduction |
| Chemistry, molecular biology, biophysics | Integration of synthetic amino acids in proteins (Xenobiology) |
| Chemistry | Design of bacteria for environmental purposes, standardization |
| Physics | Mathematics in biology, development of quantitative models |
| Microbiology | Oxygenic photosynthesis in cyanobacteria in combination with hydrogenase, development of BioBricks |
| Microbiology, biotechnology | Metabolic engineering |
| Virology | Synthetic viral and bacterial genome, minimal cells |
| Biochemistry | Protocells, origin of life |
| Astrobiology, molecular biology, evolution | Extremophile research, Synthetic biology as a tool for astrobiology |
| Biophysics | Reconstruct cellular subsystems, minimal cells, protocells |
| Biotechnology, chemical engineering | Develop molecular tools |
| Biotechnology, engineering science | Development of molecular tools |

Synthetic biologists in order of appearance: Michael Bölker, LOEWE-Centre for synthetic microbiology, Marburg (FRG); Nediljko Budisa, Institute for Chemistry, Technical University of Berlin (FRG); Victor de Lorenzo, Molecular Environmental Microbiology Laboratory, Centro Nacional de Biotecnología, Madrid (ES); Bruno Eckhardt, LOEWE-Centre for synthetic microbiology, Marburg (FRG); Bärbel Friedrich, Institute for Microbiology, Humboldt- University Berlin (FRG); Martin Fussenegger, Department of Biosystems Science and Engineering, ETH Zürich (CH); John Glass, J. Craig Venter Institute (USA); Sheref Mansy, Centre for Integrative Biology, University of Trento (I); Lynn Rothschild, NASA Synthetic Biology Earth Science Division (USA); Petra Schwille, Max-Planck-Institute for Biochemistry, München (FRG); Christina Smolke, Bioengineering Faculty, Stanford University (USA); Wilfried Weber, Center for Biological Signaling Studies, University of Freiburg (FRG). *Source* see FN1

"traditional" sub-divisions of engineering, orthogonal and protocell branches (Fig. 3.1a). It might, therefore, be helpful to analyze the means as well as the goals of synthetic biology on the basis of the disciplinary background of the protagonists (see also Bensaude-Vincent 2013 for the analysis of information technology and synthetic chemistry as roots of synthetic biology). In the search for a plausible image of synthetic biology, one could think of a mosaic picture that consists of numerous small pieces (see Fig. 3.1b), but nevertheless reveals a few overarching ideas of synthetic biology like, for example, the application of engineering principles to fundamental components of biology.[3]

---

[3]As an example of the unifying power of synthetic biology "This is a new way of thinking and it is not limited to biologists. Our science is so much more kid-driven and open to people outside of conventional biology than it's ever been before. I meet engineers; I meet physicians, people using these tools to solve problems." (John Glass, 31.05.2012) (source interviews, see FN1)

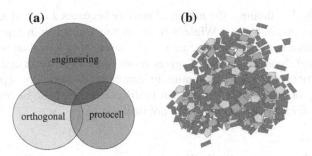

**Fig. 3.1** The three main tribes of synthetic biology (**a**) compared to a mosaic-like image of synthetic biology (**b**) that reflects the plurality of research agendas in the field caused by diverse disciplinary backgrounds of its protagonists. Researchers are closely connected to their disciplinary roots bringing along their own concepts, metaphors, tools, models and, in consequence, research agendas which contribute to the structuring and advancement of the field. This image also illustrates the lack of a clear center and dominant approach in synthetic biology

This pluralism must not be perceived as a shortcoming of the discipline, quite the opposite. The diversity of disciplines involved in synthetic biology appears to be not a temporary phenomenon typical of young disciplines, but rather a characteristic attribute of synthetic biology. In many interviews[4] the interdisciplinary character of synthetic biology was emphasized as a value,[5]—also in the long run.[6]

In summary, the pluralisms of research agendas, the lack of a coherent group of researchers, coupled with significant variations across regions, results in a lack of a linear description of synthetic biology. The reality of the research field does not evoke an empirical and self-evident definition of synthetic biology. Also, disciplinary boundaries are still fuzzy. This situation is not necessarily a problem for the scientific advancement of the field itself and is not perceived as threatening to the discipline. However, the ambiguity of the label synthetic biology caused by the complexity of the research field is carried over to the discussion on risks associated with synthetic biology.

## 3.1.2 Dichotomy of the Attributes Novelty and Continuity

When synthetic biology is discussed in the contexts of science policy, legislation and public funding, the situation becomes even more complex. The scope of synthetic biology and its attributes are interpreted quite differently depending on the

---

[4]See FN1.

[5]For example „You get a lot of very exciting research and breakthroughs when you bring people with different disciplines together on a problem" (Lynn Rothshield, 14.06.2012); "Because if it becomes a sort of, let's say synthetic biology training—we will lose some of the interdisciplinarity, which will not be nice." (Sheref Mansy, 15.06.2012), (source interviews, see FN1).

[6]However having in mind the rapid speed of synthetic biology, the statements were cautious with respect to predictions (for positive predictions see also Schmidtke and Schmidtke, 2007.).

strategic goals. For funding, the notion of novelty becomes a central argument for promoting the research field. While novelty is emphasized with regard to funding, it could turn out to be an obstacle in the context of risk evaluation. However, the vague label "synthetic biology" gives room to an often-practiced ambiguous use of the attributes novelty and continuity. Similarly, the different aspects of synthetic biology or the continuity argument in discussing synthetic biology might be played out differently depending on the political context.

### 3.1.2.1 Novelty in the Synthetic Biology Context

Right from the beginning of synthetic biology, the narratives of novel futures and novel strategies have been central to identity building within the field and—what is even more important—have been an inherent strategic part of synthetic biology. Or, as Christina Agapakis put it, "a discipline defined much better by shared dreams for the future rather than any present technique or application."[7] Grand visions or even science fiction are not seen as pure phantasies, but rather as thought experiments that help to understand the needs of the present world. This "design fiction"[8] helps to develop projects that start with a simple "what if…?". One example of far reaching visions can be found in the book Regenesis of George Church, one of the leaders in the field, about advances in genome editing that may make it possible to clone extinct species, including Neanderthals. But the talk on novel futures is also aiming at creating prototypes that convince others (in particular venture capitalists and funders) to see the potential of synthetic biology.[9] However, this twofold role of helping the field to develop scientific roadmaps and promoting the field in the public can turn out to be a double-edged, since in the political realm novelty is often connected with novel risks.

The example of Drew Endy, another forerunner in the field, shows some adaptation to these political realities. He was selling visionary programs for the future of synthetic biology, emphasizing its novelty and revolutionary status. A couple of years later, after having closer contact with policy makers for whom such attributes might have raised concerns, he switched from a more revolutionary rhetoric to a strategic downplaying of the novelty and risks of the new field (Campos 2013). The dichotomous use of novelty—emphasizing the novelty in the context of funding while downplaying it in the frame of risk discussion to minimize legal hurdles—was (and sometimes still is) also common in genetic engineering communication (Tait 2009).

As long as the term novelty is used in the described ambiguous way debates about synthetic biology get caught in the gap between the existing realities of the field and

---

[7]Synthetic Biology as Collective Fantasy by Christina Agapakis, http://studiolab.di.rca.ac.uk/blog/synthetic-biology-as-collective-fantasy, assessed 11 August 2015.

[8]See FN above.

[9]See FN above.

design fiction. One way out of this dilemma could be to start a discussion in the political realm, whether novelty is at all a good indicator of novel risks. Novelty has a clear normative level with references to context and purpose (Guston and Brian 2015) but is consequently a rather unreliable indicator of potential risk. It misses out the fact that conventional risks are not less risky, only because we are used to them, but can just as well lead to harm. Some of these risks will be new and we will have to develop tools to handle them. But the same is true for conventional risks that are not obvious. In all cases though, it will be essential to identify and manage them.[10]

### 3.1.2.2 Invalidity of the "Continuity Argument" in the Context of Risk Evaluation of Synthetic Biology

While the notion of novelty plays its role in the political realm *and* within the field of synthetic biology, the continuity argument is almost exclusively been found in political contexts. It appears in two different forms: either the continuity of synthetic biology to "thousand of years of experience in breeding" is spelled out or the continuity between genetic engineering and synthetic biology is emphasized.

Whether or not it makes sense to refer to such continuities, however, depends on the context in which synthetic biology is discussed. For example in the context of science history it is quite adequate to emphasis continuities, whereas in the frame of risk evaluation it is important to focus on discontinuities to identify fields where action is required. The reference to traditional breeding in risk assessment of new applications of biotechnology leads to the false and potential dangerous conclusion, that no new risk evaluation of the technology is needed. It is quite tempting to assume that old practices can serve as a carte blanche for future applications. However as Richard Twine unmasks the argument in the context of transgenic animals it is an "ethical bypass" (Twine 2010, p. 54). Even if traditional breeding, modern genetics, and biotechnological methods all aim to optimize organisms for the good of man, they differ on the level of risk evaluation. It is quite questionable, whether the traditional praxis to breed plants, animals and microbes also legitimate deeper, more targeted and quicker interventions in the germ line. The evolutionary space is already expanded in genetic engineering, when species barriers are getting blurred. In synthetic biology the evolutionary realm does not act as a limiting factor any more and thus disconnects applications more and more from existing organisms. This higher depth of intervention is the most important difference to traditional breeding (see Sect. 3.3.1). It leads to a loss of familiarity to known, natural organisms, which we could refer back to. However current risk assessment systems rely on the familiarity-principle that is based on the knowledge on behaviour and risk profiles of organisms experienced and collected in centuries. With the increasing loss of familiarity, this traditional principle for risk assessment has to be increasingly suspended. The continuity argument

---

[10]http://2020science.org/2015/02/02/novelty-nanomaterials-overrated-comes-risk/#1.

in the context of risk assessment seeks to convey that we still dispose over such experienced knowledge also in the context of new applications of biotechnology. Sometimes even a transmission of public acceptance of traditional breeding technologies to the modern biotechnology is hoped to be achieved by emphasising the continuity to traditional practices. The first part of the continuity argument, which places synthetic biology in a wider historic and technological connection is surely correct. Due to time limits and evolutionary constraints, however, the depth of intervention in organisms is limited in traditional breeding. Acceleration of speed is characteristic for modern biotechnology starting from marker assisted breeding, over genetic engineering to synthetic biology. In combination with the vanishing limits of evolutionary provisions this leads to a rising depth of intervention in organisms and a loss of the familiarity principle. Thus the second part of the continuity and breeding argument that suggests that applications of modern biotechnology and synthetic biology are automatically safe because of their historic background is insufficient and illegitimate.

### 3.1.3 Scope of the Synthetic Biology Politicized

Besides the already described obstacles to grasp the reality of the research field synthetic biology, in the discussion on risk evaluation and funding the scope of the label "synthetic biology" can in addition become politicized. When it comes to funding policies and promotion of the field, very wide definitions of synthetic biology are regularly used to promote the field. In contrast, very narrow definitions are common when regulations are discussed.[11] Definitions of synthetic biology have therefore often a political dimension.

### 3.1.4 Time Point and Speed of the Scientific Development Differentiated

Another important aspect that needs to be taken into account in differentiating the discussion on risks is the time point of the scientific development the discussion is on. For the majority of current and near-term commercial and industrial

---

[11]See for example the Convention on Biological Diversity online discussion about synthetic biology, Topic 3: "Operational definition of synthetic biology, comprising inclusion and exclusion criteria", where operational definitions are discussed, bch.cbd.int/synbio/open-ended/pastdiscuss ions.shtml#topic3, Accessed 19 June 2015. Jim Thomas from the ETC-group (post [#6829]), for example, suggests a very wide definition, whereas Steven Evans from Dow AgroSciences (post [#6877]) writes that "one line in the sand for separating 'traditional' molecular biology and synthetic biology is the point at which the resulting organism, irrespective of how they were inspired or how they were actualized, can no longer exchange information or transcribe/translate information with its originating species strain or any other 'natural' species." Thus, in effect, Evans suggests restricting an operational definition of synthetic biology to xenobiology.

applications of synthetic biology the risk assessment framework is often being discussed as sufficient in most cases though it has its pitfalls already for existing applications in genetic engineering. With respect to future applications it needs to be assessed at what point our current instruments are no longer fully adequate. This will especially be the case when the principle of familiarity cannot be applied for the risk assessment procedure anymore. In current risk assessment discourses a wide range in viewpoints both on the status on current synthetic biology applications and on the expectations for future synthetic biology applications exist. In calling for a sound discussion on synthetic biology it has been emphasized neither to hype the hopes nor the risks of synthetic biology (Sauter 2011; Kaiser 2012). However if we are too cautious in the discussion we might underestimate the field. Even though synthetic biology has not yet succeeded to synthesize a single living cell from the scratch, nor has it brought back the mammoth to live, one should not underestimate the rapid development within synthetic biology that has already been accomplished. As the case of genome editing already shows, some developments may even occur sooner than anticipated or in a much expanded fashion. Especially the CRISPR/Cas9 system can act as a catalyst for synthetic biology by accelerating the speed and extending the scope of synthetic biology—moving away from a limited number of prokaryotic model organisms, to a wider range of organisms up to the level of higher organisms like plants and animals.

## 3.2  Level of the Risk Discussion

In addition to the above-mentioned obstacles in the risk evaluation on the label, scope and historic placement of synthetic biology, the discussion is sometimes also prone to misunderstandings, when different levels of synthetic biology are discussed in parallel. It is in a way self evident that for example synthetic biology does not pose a risk by itself, but rather the individual organisms, genetic devices (like for example the release of gene-drivers) or products (as for example xenobiological compounds), that derive from these organisms. Next to it risk evaluation needs not only to evaluate the nature of the particular genetic changes but also the specific use of that organism. On this level the biggest difference lays in the exposure routes to environment and humans: is the application designed for contained use, field release or gene therapy? This has direct consequences for the assessment of potential risks for biodiversity and human health. Especially in cases with a great level of uncertainty that we discuss below, the retrievability and recovery of the organisms is of great importance. Most organisms (except some sterile organisms or large animals) can however no longer be retrieved once released. Most of current and near future synthetic biology applications are aimed for contained use, but the volumes envisioned are rising and by this also the probability of unintentional release. Thus questions on the policy of containment practices will supposedly gain momentum.

The described further structuring is of course somehow trivial but nevertheless in praxis again and again underrepresented. Thus awareness of these different levels in discourse could be a helpful tool.

## 3.3 New Features of Synthetic Biology Relevant for Risk Evaluation

As elaborated in Sect. 3.1 on the level of risks evaluations the ambiguity in the use of the label "synthetic biology" in the political context can become an obstacle for a differentiated discussion on potential risks of synthetic biology applications. We thus suggest focusing not on the label "synthetic biology", but on the *features* that synthetic biology brings along and that are relevant to the societal evaluation. To approach the risk discussion it is important to identify critical features that are new to synthetic biology or its subfields when compared to genetic engineering and that are important for an adapted risk evaluation. These new features are in particular the growing depth of intervention in the organism and, the associated decreasing familiarity of synthetic organisms that together with the high speed of technological development challenge established risk assessment systems. This should however not imply that other risk assessment categories that are already valid for genetic engineering are not significant as well.

### 3.3.1 Depth of Intervention, Complexity and Predictability

#### 3.3.1.1 Depth of Intervention

Although synthetic biology builds on many of the techniques of genetic engineering, it aims at the design and construction of new biological parts, devices and systems. Thus, it goes far beyond mere modification of existing cells by inserting or deleting single or a few genes. Instead, cells will be equipped with new functional devices or even complete biological pathways and systems will be designed. Therefore, in comparison to genetic engineering, organisms produced by synthetic biology are characterized by a much larger depth of intervention into the biological nature of the organism,—both by quality and quantity. The quality level is mainly introduced by the new sequence space that synthetic biology can explore by designing novel genomes beyond evolutionary realm (compare layer model, Fig. 2.1 of Chap. 2). The resulting synthetic organisms may have only very little in common with any other organism and predictions on the behavior of these synthetic cells cannot simply be drawn from the behavior of any of the donor organisms that have been used for the creation of synthetic cells. Another dimension of the depth of intervention is the introduction of unnatural chemical compounds into living systems. This results in cells, whose metabolism or genetic material is incompatible with that of natural biological systems. Although this orthogonality of xeno-organisms is also discussed as a means to enhance biosafety, such organisms may have a yet unpredictable impact on the ecosystem if released into the environment. This comes along with a lack of detection methodology for xeno-organisms (see also Sect. 3.4).

### 3.3.1.2  Predictability: Safety Concepts of Synthetic Biology

Synthetic biology as engineering technology based on living systems may claim that reduction of complexity is the only way to create living cells with a completely predictable behavior. However, it is foreseeable, that cells engineered for specific purposes have to be equipped with additional functions to enhance their stability and robustness. It is well known that even in classical engineering technologies the construction of ever more complex systems is accompanied with an increase of inherent instability and uncertainty. Even for mathematical systems it was proven that they will contain theorems that cannot be decided. Thus we are left with a level of uncertainty even in a world of complete predictability. In synthetic biology as generalized strategy it is commonly accepted that the design and construction of novel biological functions works best on the basis of a minimalized cell. Such minimal cells can be used as chassis to set up artificially designed cells that can fulfil useful functions that are novel and have not been seen in nature before. Estimation of parameters, predictability, reducing of complexity by standardization and modularization, aim is the total control over the behaviour of the "system" This concept results from the world view: life is composed of devices and modules and can be described as an integrated complex system (Andrianantoandro et al. 2006) like a computer which is built from simpler components like cards, processing units, transistors etc. This engineering view allows the construction of complex systems by using a highly effective 'hierarchy' of communication between engineers of different system levels that each are responsible and (perfect) in their field.

Risk comes here not only from the inherent unpredictability of living organisms that might be retained in spite of all engineering, but also form the painful experience that has been made with recent technologies dealing with highly complex technologies like personal computers, large electric power transmission grids and nuclear power plants. While personal computers often crash and just have to rebooted, failure of the latter can result in nation-wide blackouts or may cause high financial and material losses and even make (large) areas uninhabitable. In all these cases it is the large complexity of these highly integrated modular systems that inherently bears the risk of unpredictable behaviour.

While genetically modified organisms, in which single or only few genes have been manipulated or been introduced may be regarded as safe, the high complexity of organisms carrying diverse genes of different origin or even designed genes with no natural counterparts, may carry a similar risk as highly complex technical systems. This not necessarily enhances the actual risk in terms of potential damage or danger but results in an unpredictability, which enhances the uncertainty connected with these artificial cells.

In contrast to many technical systems where risk assessments can be made more or less precisely (even for worst case scenarios) this appears difficult for synthetic biology. The potential damage (if any) is hard to estimate and at the same time the probability of occurrence is nearly indeterminable. Therefore these systems are afflicted rather with uncertainty than with risk

## 3.3.2 Risk Assessment Challenged: Familiarity Principle and Speed of Developments

Closely connected to the rising depth of intervention are challenges to current comparative risk assessment. Especially in those areas of synthetic biology that go beyond genetic engineering by broadening the evolutionary realm suitable 'comparators' are missing due to a growing lack of familiarity to natural organisms. Particular in those cases our ability to predict potential risks to humans and biodiversity is limited by a lack of experience with comparable cases. We can gain some experience with research and a step-by-step approach, but in the cases of low similarity and familiarity to natural organisms, even this approach can bring along very limited knowledge within a relevant timespan. Therefore, the traditional assessment of potential risks according to the hazard potential of both donor and acceptor organism reaches theoretical and technical limits. Even beyond these limits will be the estimation of risks connected with cells and organisms that have been designed from the scratch on computers or that have no natural counterparts. These problems are further complicated, with the rising speed of developments—another relevant feature of synthetic biology. The number of introduced transgenic traits and types of donor and acceptor organism increases, which—in combination with the sheer volume and speed of introduction of novel organisms—will bring a 'case by case' evaluation to its limits. Thus the comparative risk assessment system based on the familiarity principle and case-to-case approach will be at stalk as the production systems for novel organisms become more and more sophisticated and effective. Risk assessment mechanisms will have to be adapted in parallel, taking into account the whole novel organisms as the depth of intervention increases and the limitations in the time frame in a step-by-step approach. To prevent acting under uncertainty with respect to risks to humans and biodiversity in the cases of low familiarity to natural organisms, the implementation of the precautionary principle and in addition an ongoing dialog between scientists and society is of special importance.

## 3.4 Special Cases of Risks Assessment in Synthetic Biology

With respect to orthogonal biology and protocell research some special aspects are discussed in the following. For all layers that contain orthogonal organisms the current lack of knowledge on potential interactions with the biotic and abiotic environment makes any established risk assessment methodology nearly impossible. Xenobiological life forms that do not rely on DNA as their genetic base are unfamiliar both for us and the environment and thus challenge not only the traditional methods of risk analysis but also pose questions to their artificial nature and to the concept of life.

## 3.4.1 "The Farther, the Safer"?

Genetic encoding in living systems is based on a highly standardized chemistry composed of the same number of 4 nucleotides or "letters" as building blocks of the informational polymers DNA and RNA, respectively, and the twenty α-amino acids as basic building blocks for proteins. The universality of the genetic code enables the horizontal transfer of genes across biological taxa, which afford a high degree of standardization and interconnectivity. Thus, all changes that affect this chemistry within living systems tend to be generally lethal (Marlière 2009; Schmidt 2011; Acevedo-Rocha and Budisa 2011). In this context, one of the great challenges is the development of a strategy for expanding the standard basic chemical repertoire of living cells to inter alia improve biocontainment by man-made or naturally evolved changes in the genetic code (Mandell et al. 2015; Rovner et al. 2015). All basic constituents of DNA (the heterocylic bases, the deoxyribose sugar and the connecting phosphodiester backbone) are exchangeable with alternative chemical structures such as novel base pairs (Piccirilli et al. 1990) or xeno-DNA nucleic acids (Pinheiro et al. 2012). On the other hand, long-term cultivation experiments with quick-growing asexual bacterial cells, as pioneered by Lenski (Blount et al. 2012) with serial dilutions using conventional chemostats or turbidostats, also showed notable success. Thus, the canonical chemical barrier was recently surpassed by evolving an *E. coli* strain with a genome in which thymine was replaced 5-chlorouracil (Marlière et al. 2011). This is a first important example in the construction of biological systems composed of more than 90 % thymine replaced with 5-chlorouracil, that are thus composed of xeno-nucleic acids, to a large extent.

One of the arguments to follow this path of orthogonality is the safety aspect. "The farther, the safer" is the slogan of orthogonality (Marliere 2009). The idea behind this approach is to create synthetic organisms that keep all the complexity and unpredictability inherent to life but to control these artificial life forms by efficiently preventing any genetic interaction with natural cells. Essentially, these cells are in total genetic isolation from extant species for an indefinite duration. Thereby unanticipated forms of life could emerge from evolving populations with non-canonical xeno-proteins and nucleic acids making plausible the advent of a parallel xeno-biological world (Schmidt 2010; Budisa 2014).

Orthogonal artificial living beings are thus separated from nature by a genetic fence or "firewall". Although this approach might be theoretically tight it leaves many observers with the same feeling as if wild animals are watched in a zoo behind a glass window or a moat: "What happens, if…?" Again we are left with a sense of uncertainty although we cannot explain by which means the fail-safe system might nevertheless flop, at least with respect to genetic exchange between xeno-organisms and natural organisms. Even if orthogonalization precludes any unwanted spreading of genetic information into the environment one has still to consider both the ecological impact at the level of competition for nutrients and toxicological aspects of the xeno-material used to construct these cells. Not only is the genetic material incompatible with that of natural biological systems, xeno-biological organisms differ also in their metabolism. The release of e.g. xeno-DNA (XNA) into the environment may have a yet unknown impact on the ecosystem. Many of these molecules are chemical compounds that are new-to-nature and thus may possibly accumulate in the environment. It is known that xenobases such as chloruracil are stable during autoclave sterilization (Patra et al. 2013). There exists no established legal regime for these alien compounds since REACH is not applicable for such small quantities. Furthermore, large-scale propagation of orthogonal cells or introduction of unnatural chemical compounds into living systems requires the development of appropriate detection methods for these cells and compounds in environmental settings. In addition, traditional methods like DNA-amplification by PCR that are extremely sensitive with naturally occurring DNA may not work to detect xeno-DNA and/or xenobiological cells in the environment.

### 3.4.2 Sense Codon Reassignment in the Genetic Code and Its Possible Consequences

Some of these problems might not apply to organisms in which reassignment of one or several codons of the genetic code serves as strategy for orthogonalization. Some radically new technologies and materials in future will be generated by non-canonical amino acids (ncAAs)[12] or their precursors that are expected to be the key xenobiotic building blocks (Bohlke and Budisa 2014; Liu and Schultz 2010). Via newly designed biosynthetic pathways, the cellular metabolism will be redirected to produce ncAAs. Combined with novel orthogonal pairs (o-pairs) of tRNA and aminoacyl-tRNA synthetase, peptides and proteins will be endowed with these artificial chemical handles—expanding the genetic code and introducing xenobiotics into the chemistry of life. Thereby, it is not difficult to anticipate the power of orthogonalization to engineer complex biological systems in order to (a) provide solutions of industrially relevant bio-production problems, such as peptide and protein production beyond the canonical set of natural molecules; (b) to expand the arsenal of chemistries available for living cells (Budisa 2014).

This is indeed plausible, since it is now well established that between 30 and 40 sense codons in the genetic code are adequate to encode the genetic information of

---

[12]Noncanonical amino acids, are amino acids that do not occur in the genetic code.

an organism (O'Donoghue et al. 2013), a large number of sense codons (>20) may be available for recoding with ncAAs. Current genetic code expansion approaches (e.g. Chin 2014), programmed to reassign UAG or UGA stop codons, are geared to produce new proteins or biomaterials containing one or multiple ncAAs with fluorescent or chemically reactive groups for a host of in vivo or in vitro applications. However, proteome-wide replacements of amino acids by sense codon reassignments are not practical by this route as recently demonstrated by long-term *E. coli* evolution experiment that resulted in an organism in which all 20,899 UGG codons were 'recoded' by substituting the supply of the original tryptophan with thienopyrrolyl-alanine (Hoesl et al. 2015). This example of a proteome-wide amino acid replacement represents a significant 'genetic code expansion' and a real step towards a synthetic organism. Ultimately, the progressive reprogramming of existing biological functions will lead to artificial life ideally so isolated from natural life that surviving outside the laboratory is not possible.

The strategy of codon reassignment differs with respect to orthogonalization, from the earlier described case of xenobases (see Sect. 3.4.1). In the case of codon reassignment the natural chemical makeup of cellular and genetic metabolism is maintained. Physical exchange of genetic material (DNA) between these cells and their natural counterparts might still occur. The alternative genetic code, however, prevents any useful exchange of genetic information since the incoming DNA will be translated into a completely different protein. In this scenario the genetic firewall relies on semantic differences rather than on grammatical ones.

### 3.4.2.1 Natural Background

It is generally accepted that the genetic code has developed from earlier life forms with fewer amino acids. Its current structure is considered to be evolved from a 'frozen accident' during early development of complex life on Earth (Gusev and Schulze-Makuch 2004; Sella and Ardell 2006). Thus, the genetic code is the oldest molecular fossil descending directly from the last-common ancestor—a grantee of the biochemical uniformity in all living organisms. At the time of its deciphering, the genetic code was suggested to be universal in all organisms. Crick proposed a first theory on its evolution saying that the genetic code was a result of a "frozen accident" unable to evolve further, since "no new amino acid could be introduced without disrupting too many proteins" (Crick 1968; Söll and RajBhandary 2006).

However, the theory of a frozen universal genetic code was disproved by genetic code variations found in vertebrate mitochondria, where AUA encodes Met instead of Ile and UGA encodes Trp instead of being a translational termination signal (reviewed in: Budisa 2005). Today, around 20 variations of the standard genetic code are known, not only in organelles but also in free-living microbes (O'Donoghue et al. 2013).

Changes in codon meaning can be classified into two general categories: the first of such variations includes the naturally occurring reassignment of standard stop codons (UGA, UAA and UAG) to Trp or Gln found in some prokaryotes and archaea, but also in mitochondrial genomes and even in nuclear genes of some protozoa (e.g. *Tetrahymena* and *Paramecium*). The second category includes altered meanings in mitochondrial sense codons such as Met → Ile; Lys → Asn; Arg → Ser and Leu → Thr. The only documented case of natural sense codon reassignment among nuclear genes is the incorporation of serine instead of leucine at CUG codons in the *Candida, Debaryomyces,* and *Lodderomyces* genera (CTG clade) of ascomycetous fungi (Butler et al. 2009). However, it should be kept in mind that these reassignments do not affect the standard (canonical) amino acid repertoire. For that reason, Bohlke and Budisa reasoned such reassignments might be seen as flexibility in the "frozen" structure of the genetic code as no known examples for codon reassignment introduce a non-canonical amino acid in response to a sense codon (Bohlke and Budisa 2014).

### 3.4.2.2 Adaptive Features

From the experimentalist's point of view, the departures in codon reassignment although described as "evidences" for the 'continuous' evolution of the genetic code, are in fact functional adaptations. For example, the well-documented (CUG) Leu → Ser reassignment in some genes of *C. cylindrica* is a useful adaptation to stress response (Santos et al. 2004). In particular, the CUG: Leu → Ser change would be expected to exert a strong negative impact on the organism's fitness. The reassignment of CUG in phylogenetically related *Saccharomyces cerevisiae* by introduction of tRNACAG gene from *C. cylindrica* yielded in yeast strain with harmed viability due to the production of aberrant proteins. However, in *C. cylindrica* this reassignment is 'reserved' only for a tiny portion of genes involved in stress response and pathogenicity; the CUG triplet does not appear in most of the cellular mRNA.

### 3.4.2.3 Technological Risk and Benefits

Experimental attempts to develop heterologous expression systems with an extended amino acid repertoire change at the level of single target proteins were successful in the last decades (Liu and Schultz 2010). These experiments should enable the rebuilding of the code structure based on a different set of chemicals as the current one and will yield tailor-made proteins and designer cells with novel teleonomic determinants; these can bring about new technological promises: environmental-friendly biotechnologies, novel tools in diagnostics as well as in fighting infection diseases, to mention but a few. However, such developments also might bring about potential risks as well. For example, it is conceivable to generate lethal viral/bacterial/fungal strains with their 'own' genetic codes resistant to all known defense repertoires. In addition it might be capable to introduce lethal sense-to-sense or sense-to-nonsense codon reassignments into their hosts and subsequently seriously harm or even kill them either in selective or non-selective manner. In nature such detrimental codon reassignment (see above) is already well documented at least in one case.

### 3.4.3 The Innocence of Protocells

In general, the protocell approach appears to be the least risky one. This is mainly due to the inherent simplicity of protocells that are being generated. Although nature has been quite successful in evolving living systems of unanticipated complexity "from the scratch", this process has taken some time (at least 3 billion years) and started under completely different environmental conditions. The pristine soup contained many energy-rich compounds in a highly reducing atmosphere. If released into the current environment, protocells would have to compete with billions of perfectly adapted microorganisms. Therefore, survival of protocells is conceivable only in completely controlled environments. Escape from these closed compartments might be imaginable but it is hard to believe that such cells have the tiniest chance to compete with the omnipresent and well-adapted natural organisms. It may be thus no wonder that ethical considerations regarding protocells so far rather deal with the "ethics of protocells" asking the questions whether these organisms are already alive or not and whether they contribute to biodiversity thus have to be protected.

## 3.5 Summary and Outlook

This chapter aims to offer some tools to the discussion on possible risks that might be elicited by synthetic biology, by describing areas of misunderstandings, connected to the label and scope of synthetic biology and the novelty and continuity argument. With the focus on specific features of synthetic biology that are relevant to the risk discussion, some areas of discourse on suitable risks assessment methodology might be facilitated. However this will only cover the strand of primary risk assessment. Secondary effects of synthetic biology applications such as socioeconomically effects, as for example the impact to small-scale farming, will clearly add up to the already complex situation in synthetic biology risk assessment. Here, it becomes even more important to approach these challenges in a multi-disciplinary way—and also in line with the precautionary principle

# References

Andrianantoandro E, Basu S, Karig DK, Weiss R (2006) Synthetic biology: new engineering rules for an emerging discipline. Mol Syst Biol 2:28

Acevedo-Rocha CG, Budisa N (2011) On the road towards chemically modified organisms endowed with a genetic firewall. Angew Chem Int Ed Engl 50:6960–6962

Bensaude-Vincent B (2013) Discipline-building in synthetic biology. Stud Hist Philos Biol Biomed Sci 44:122–129

Blount ZD, Barrick JE, Davidson CJ, Lenski RE (2012) Genomic analysis of a key innovation in an experimental *Escherichia coli* population. Nature 489:513–518

Bohlke N, Budisa N (2014) Sense codon emancipation for proteome-wide incorporation of non-canonical amino acids: rare isoleucine codon AUA as a target for genetic code expansion. FEMS Microbiol Lett 351:133–144

Budisa N (2005) Engineering the genetic code—expanding the amino acid repertoire for the design of novel proteins. WILEY-VHC Weinheim, New York

Budisa N (2014) Xenobiology, new-to-nature synthetic cells and genetic firewall. Curr Org Chem 18:936–943

Butler G, Rasmussen MD, Lin MF, Santos MAS, Sakthikumar S, Munro CA, Rheinbay E, Grabherr M, Forche A, Reedy JL, Agrafioti I, Arnaud MB, Bates S, Brown AJP, Brunke S, Costanzo MC, Fitzpatrick DA, de Groot PWJ, Harris D, Hoyer LL, Hube B, Klis FM, Kodira C, Lennard N, Logue ME, Martin R, Neiman AM, Nikolaou E, Quail MA, Quinn J, Santos MC, Schmitzberger FF, Sherlock G, Shah P, Silverstein KAT, Skrzypek MS, Soll D, Staggs R, Stansfield I, Stumpf MPH, Sudbery PE, Srikantha T, Zeng Q, Berman J, Berriman M, Heitman J, Gow NAR, Lorenz MC, Birren BW, Kellis M, Cuomo CA (2009) Evolution of pathogenicity and sexual reproduction in eight Candida genomes. Nature 459:657–662

Campos L (2013) Outsiders and in- laws: drew endy and the case of synthetic biology. In: Harman O, Dietrich MR (eds) Outsider scientists. Routes to innovation in biology. University of Chicago Press, Chicago, p 339

Crick FHC (1968) The origin of the genetic code. J Mol Biol 38:367–379

Chin JW (2014) Expanding and reprogramming the genetic code of cells and animals. Annu Rev Biochem 83:379–408

Dormitzer PR, Suphaphiphat P, Gibson DG, Wentworth DE, Stockwell TB, Algire MA et al. (2013) Synthetic generation of influenza vaccine viruses for rapid response to pandemics. Sci Transl Med 5:185ra68

Engelhard M, Hagen K (2012) Bedeutung der Grenzüberschreitung zwischen Disziplinen und Akteuren in der Synthetischen Biologie, Non public expertise conducted for the German parliament

Gibson DG, Glass JI, Lartigue C, Noskov VN, Chuang R-Y, Algire MA et al. (2010) Creation of a bacterial cell controlled by a chemically synthesized genome. Science 329:52–56

Glass JI (2012) Synthetic genomics and the construction of a synthetic bacterial cell. Perspect Biol Med 55:473–489

Gusev VA, Schulze-Makuch D (2004) Genetic code: lucky chance or fundamental law of nature? Phys Life Rev 1:202–229

Guston DH, Brian JD (2015) Workshop on the research agenda in the societal aspects of synthetic biology. Talk. http://cspo.org/wp-content/uploads/2015/01/NewTools_Talk_24Apr2015_final.pdf Accessed 14. Nov 2015

Hoesl MG, Oehm S, Durkin P, Darmon E, Peil L, Aerni H-R, Rappsilber J, Rinehart J, Leach D, Söll D, Budisa N (2015) Chemical evolution of a bacterial proteome. Angew Chem Int Ed Engl 54:10030–10034

Kaiser M (2012) Commentary: looking for conflict and finding none? Public Underst Sci 21:188–194

Knuuttila T, Loettgers A (2014) Varieties of noise: analogical reasoning in synthetic biology. Stud Hist Philos Sci Part A 48:76–88

Liu CC, Schultz PG (2010) Adding new chemistries to the genetic code. Annu Rev Biochem 79:413–444

Mandell DJ, Lajoie MJ, Mee MT, Takeuchi R, Kuznetsov G, Norville JE, Gregg CJ, Stoddard BL, Church GM (2015) Biocontainment of genetically modified organisms by synthetic protein design. Nature 518:55–60

Marlière P (2009) The farther, the safer: a manifesto for securely navigating synthetic species away from the old living world. Syst Synth Biol 3:77–84

Marlière P, Patrouix J, Döring V, Herdewijn P, Tricot S, Cruveiller S, Bouzon M, Mutzel R (2011) Chemical evolution of a bacterium's genome. Angew Chem Int Ed Engl 50:7109–7114

O'Donoghue P, Ling J, Wang YS, Söll D (2013) Upgrading protein synthesis for syntheticbiology. Nat Chem Biol 9:594–598

Patra A, Harp J, Pallan PS, Zhao L, Abramov M, Herdewijn P, Egli M (2013) Structure, stability and function of 5-chlorouracil modified A:U and G:U base pairs. Nucleic Acids Res 41:2689–2697

Piccirilli JA, Krauch T, Moroney SE, Benner SA (1990) Enzymatic incorporation of a new base pair into DNA and RNA extends the genetic alphabet. Nature 343:33–37

Pinheiro VB, Taylor AI, Cozens C, Abramov M, Renders M, Zhang S, Chaput JC, Wengel J, Peak-Chew S-Y, McLaughlin SH, Herdewijn P, Holliger P (2012) Synthetic genetic polymers capable of heredity and evolution. Science 336:341–344

Rovner AJ, Haimovich AD, Katz SR, Li Z, Grome MW, Gassaway BM, Amiram M, Patel JR, Gallagher RR, Rinehart J, Isaacs FJ (2015) Recoded organisms engineered to depend on synthetic amino acids. Nature 518:89–93

Santos MAS, Moura G, Massey SE, Tuite MF (2004) Driving change: the evolution of alternative genetic codes. Trends Genet 20:95–102

Sauter A (2011) Synthetic biology: Final technologisation of life—or no news at all? TAB-Brief 39 Special Edition, http://www.tab-beim-bundestag.de/en/research/u9800/TAB-Brief-39-EN-SynBio.pdf Assessed 2 Aug 2015

Schmidt M (2010) Xenobiology: a new form of life as the ultimate biosafety tool. BioEssays 32:322–331

Sella G, Ardell DH (2006) The coevolution of genes and genetic codes: Crick's frozen accident revisited. J Mol Evol 63:297–313

Söll D, RajBhandary UL (2006) The genetic code—Thawing the 'frozen accident. J Biosci 31:459–463

Suzuki Y, Assad-garcia N, Kostylev M, Noskov VN, Wise KS, Karas BJ et al. (2015) Bacterial genome reduction using the progressive clustering of deletions via yeast sexual cycling. Genome Res 1–11

Tait J (2009) Governing synthetic biology: processes and outcomes. In: Schmidt M, Keller A, Ganguli-Mitra A, de Vriend H (eds) Synthetic biology. The technosciences and ist social consequences. Springer, Berlin, p 150

Twine R (2010) Animals as biotechnology: "Ethics, sustainability and critical animal studies". Earthscan, London, p 54

# Chapter 4
# The Concept of Life in Synthetic Biology

Georg Toepfer

**Abstract** In this chapter I analyze the research program of synthetic biology from a conceptual point of view. With this aim, I first highlight four episodes in the history of artificial life research. To give a more abstract characterization of the synthetic aspects of artificial life, I further propose to differentiate between eight categories of artificiality arranged in two broad groups, the first starting from living and the second from non-living precursors. I then discuss artificial systems as models for living systems in general, by pointing out the conceptual affinities in our description of organisms and machines as two types of functional systems that are clearly separated from inorganic bodies. However, despite the fact that living beings are considered as a separate and distinct class, the concept of life is very ambiguous. This is illustrated by giving a list of ontological categories to which the referent of "life" may belong. In conclusion, I discuss whether artificially designed systems can be instantiations of life at all and I point to the influence the potential success of synthetic biology may have on our concept of life by strengthening its individualistic aspects and neglecting, or, for reasons of safety and sustainable use, even intentionally suppressing its evolutionary side. At any rate, the success of synthetic biology will remove one central unifying link of all forms of life on earth at the moment, common descent.

"Life" is a common term in our everyday language which is used both for the designation of our own being in the world and for the description of other beings in nature. Because of its status as an important reflexive term it has value-laden connotations but, at the same time, it serves as a merely descriptive concept within natural science. One common aspect of the two meanings, the reflexive and the descriptive, is the opaque character and self-referential structure of those entities considered as "living". Due to these attributes, living beings are often conceptualized in contrast to completely understood systems with an entirely predictable

G. Toepfer (✉)
Zentrum für Literatur- und Kulturforschung, Schützenstraße 18, 10117 Berlin, Germany
e-mail: toepfer@zfl-berlin.org

© Springer International Publishing Switzerland 2016        71
M. Engelhard (ed.), *Synthetic Biology Analysed*, Ethics of Science
and Technology Assessment 44, DOI 10.1007/978-3-319-25145-5_4

behaviour. As many man-made machines are of this type, the opposition of "living beings" and machines has emerged. This distinction can best be made clear in teleological terms: whereas machines serve external functions for the use of humans, living beings are characterized by an internal teleology, because all their activities are aimed at their own preservation and propagation. This idea leads back to Kant's concept of organisms as "natural purposes" or "self-organizing beings" in nature, and to Goethe, who imagined "the closed animal as a little world that exists for itself and through itself" such that every living being is a "purpose for itself" (Goethe 1795, p. 125). In the nineteenth century these ideas were taken up by many authors, for example by Hegel, who described a living being as a "totality" which is not determined and alterable from outside, but which forms itself from within such that it is related to itself as a subjective unity and purpose for itself (Hegel 1829, p. 165).

With these ideas of living beings as essentially self-organizing and self-determining entities in mind, it seems to be a contradiction in terms to think of artificially designed living beings. It is part of our traditional concept of life that living beings are causally autonomous and functionally closed: they are produced by other systems of their kind and all their activities are directed towards their own maintenance, which may mean self-maintenance as an individual organized being or maintenance of their kind by reproduction. In this double respect an artefact could never be said to be alive, because (1) it is not produced as a result of a self-referential dynamic but as the intentional product of a producer that is of a different type than that which he produced, and (2) it is designed (as a technical product) for other purposes than its mere maintenance (for example, for producing food or medicine for human purposes). Above all, "life" is a term that we use for describing our own existence, which appears to us as not fully controllable and projectable. A strong intuition contradicts the conception of a living being which was made for predictable purposes other than its maintenance. Or, as Paul Valéry has put it, if life had a goal, it would no longer be life ("Si la vie avait un but, elle ne serait plus la vie"; 1910, p. 57). With this conceptual tradition of "life" in mind, synthetic biology could be conceived of as a necessarily failing, tragic approach for purely conceptual reasons: because of its property of being made for some purpose external to it (even if this is only to imitate natural living beings), any organism intentionally made by humans could never be a living being.

But it seems too easy to accept purely conceptual reasons for the impossibility of synthetic biology. Humans have always used and manipulated other living beings for their purposes—and this use has not prevented the ascription of "life" to these creatures. At the moment, synthetic biology obviously is a flourishing field of empirical investigations which may help to better understand the organization of living beings in nature. Therefore, it might be a more modest and appropriate task for philosophers to remind practitioners in the field of the history of their activity and its conceptual consequences. This seems to be necessary, as it is part of the rhetorical strategies of cutting-edge scientists to neglect their forerunners and to stress the revolutionary impetus of their research by looking forwards and not backwards. Philosophers can remind the scientific protagonists that there were forerunners in their field who faced similar conceptual difficulties. With these experiences in mind they can help to clarify the concepts and theories in use.

The first things to make clear are the semantic complexities of "life" and the non-scientific, cultural dimensions of the concept.

Understanding living beings as self-referential entities has resulted in a rich tradition which asserts that "life" is an essentially inexplicable phenomenon for which our concepts will never suffice. This claim has been made by many of the prominent thinkers about the concept of life, for example Wilhelm Dilthey, Henri Bergson, Ernst Cassirer, Adolf Portmann, and Erwin Chargaff (Toepfer 2011; tab. 166). This tradition continues into the present, for example in those authors who claim that "life" is a phenomenon that cannot be grasped by an external perspective, but can only be understood from within, from the perspective of each singular living being itself (Brenner 2012, p. 107). Following this line of reasoning, it is for conceptual reasons that "life" will never be completely understood; epistemological opacity will remain part of the semantics of the term.

It is very improbable that the rapid progress in biology, especially synthetic biology, will change this semantic situation such that "life" will gradually lose its aspect of inexplicability. This is because "life" is only to a small degree a scientific concept. It is a multifarious term that takes a central position in diverse cultural contexts and is described and explained by many different fields (Deplazes-Zemp 2012, p. 758). It is a concept with great symbolic power to which many societal concerns are attached. This is also the reason why "life" is one of the most potent terms of propaganda within science. In part, it is exactly the aura of mystery and inexplicability surrounding the term which makes it so attractive. Possibly, this attraction will be diminished to the degree that the program of biology is successful in explaining and, especially, making life.

But, certainly, this scientific progress will not destroy the holistic and normative dimensions of the term. With regard to science, synthetic biology may "bring to an end a debate about the nature of life that has lasted thousands of years" (Caplan 2010, p. 423). However, this scientific development touches only upon parts of the whole concept. For the entire concept and its public understanding, the main result of synthetic biology will probably be the introduction of "a new dimension of weirdness" (Rheinberger and Bredekamp 2012). This is because the public usage of "life" as the central integrating term of our existence will surely be somehow disturbed by the fact that scientists claim to be able to make "life". Nevertheless, our conception of "life" in its everyday biographical sense need not be seriously altered by this progress within science.

The new dimension of weirdness released by synthetic biology has much to do with the normative dimensions of the notion. "Life" in its public understanding is something intrinsically valuable or even sacred—"sanctity of life" is an idiomatic expression, not only in religious contexts. With this understanding in mind, life is seen as something which is in need of care and must not be manipulated. Therefore, some may fear that treating living beings as entities that can be replaced and constructed anew could imply a loss of care and responsibility towards naturally living beings and life per se. Scientific methodological reductionism, in explaining the phenomenon, may also be accompanied by ontological reductionism, for example in an extreme form, in denying the relevance of the vital phenomena that are central to the common usage of the term (Boldt et al. 2009, p. 50).

The amount of impact of the scientific demystification of "life" on the everyday use of the term remains a question to be answered in the future. At the moment, it seems most probable that the mechanistic understanding of "life" within science will, step-by-step, simply lose the biographical, normative and holistic dimensions of the concept that are central to its everyday usage.

In the following, I shall discuss traditional approaches to the definition of life and their application in artificial life research. I shall not go into ethical or practical issues but focus on theoretical problems. Already at this level there are many ambiguities in the project of "creating life". To start with, the very notion of "creating life" is, without further clarification, not an operational term because it depends on various metaphysical interpretations (Schummer 2011, p. 133). There is neither consensus on how to define "life" nor what to accept as a genuine "creation". With respect to life, the question whether having a metabolism is sufficient or reproduction necessary for a system to be alive, is a much debated one. With respect to creation, it is uncertain where to draw a line between breeding or genetic manipulation and genuine "creation". "Life" and "creation" are basic and general concepts that are not suitable for establishing clear-cut and universally accepted differentiations.

This chapter has five parts: the first gives an historical introduction by highlighting four episodes in the history of artificial life research, the second is about modes of artificiality in artificial life research, the third discusses artificial systems as models for living systems in general, the fourth covers modes of vitality in the artificial life research, and the last discusses whether artificially designed systems can be instantiations of life.

## 4.1 Four Episodes in the History of Artificial Life Research

There are four historical episodes in the development of the concept of artificial life that I should like to discuss here. The first of them is prehistoric, the second takes place in early modern times, the third at the beginning and the fourth at the end of the 20th century.

### Human Life as Artificial Life

According to a once common definition, artificial life is the life of human beings. The very term "artificial life" first appeared in this context. It was used by Peter Damian, a reforming monk of the 11th century, to designate the life of man in an urban environment in contrast to rural life ("vita artificiosa satis et lepida"; Damiani, Epistula 44, p. 68). At the end of the 18th century, Christoph Wilhelm

Hufeland, the famous physician, wrote: "Our artificial diet is made necessary by the fact of our artificial life" (1796/1798, p. 313). In the 19th century it was common practice to contrast the artificial life of civilized man with the "savage life" of primitive people (Anonymous 1868, p. 137). "The artificial life of civilization" became a well-established figure of speech (Anonymous 1870, p. 147).

In the context of philosophical anthropology of the first half of the 20th century, it is not only civilized man but man himself that is characterized by artificiality. In his famous book of 1928, Helmuth Plessner coined the expression "natural artificiality" for the position of man at the top of the organic world (Plessner 1928, p. 309). Seen from this perspective, the first artificial life on earth would be the first human beings. On the one hand, they were products of natural evolution but on the other hand, they initiated the process of cultural evolution which produced deterministic forces that were, to some extent, antagonistic to natural evolution and, therefore, established a self-referential framework for the development of human culture.

Now, however, we would expect an artificial living being to be intentionally designed, and not to be a link in the chain of living beings in the process of evolution. Artificial living beings are not elements in the continuous transformation of living beings that take place in evolution; in contrast they are intentionally designed art works.

## Automata Imitating Life Functions

Perhaps, therefore the first artificial life on earth is the first figurative art work. As far as is currently known this is the Venus of Hohle Fels, an Upper Paleolithic Venus figurine found in 2008 in Schelklingen, Swabia, Germany. It is dated to between 35,000 and 40,000 years ago and is the oldest undisputed example of Upper Paleolithic art and of all figurative prehistoric art (Conard 2009).

However, the ascription of artificial life to this material entity is problematic as, obviously, it is an inanimate body. Despite its inanimate nature, this piece of matter has some aspects of a living body, namely its form or shape that resembles living beings.

The point I want to make with this example is simply that every entity that is considered as an embodiment of artificial life bears no more than a few attributes of nonartificial life. Living entities that resemble, in all their aspects, natural living beings are natural living beings. In this example, it is only the external shape that is similar to natural living beings. This is particularly unsatisfying, because one of the most fundamental aspects of living beings is that they are, first of all, dynamic systems. They consist of parts that interact with each other and they exchange matter and energy with their environment in order to maintain their instable material organization. Therefore, it seems to be more attractive to view artificial *machines* as models, or even as embodiments, of natural living beings.

Mechanical or hydraulic devices that imitate living functions have been manufactured since antiquity. They were described, for example, in Homer's "Iliad". But the most influential book was the "Pneumatics" of Heron of Alexandria from the first century AD, which was used for constructing machines until early modern times. One of the most famous artificial living beings was the mechanical duck of Jacques de Vaucanson, which he constructed in 1738 (Vaucanson 1738; Chapuis and Droz 1958; Heckmann 1982).

Despite being made of copper plates, it was destroyed in a fire at the museum of Nizhny Novgorod in 1879. The duck was said to be able to swim in water, to move its wings in a duck-like manner, to utter duck-like sounds, and, most famously, to eat kernels of grain and to digest and excrete them. But, as a contemporary critic commented, the duck did not really have this ability, as the ingested food did not continue down the neck and into the stomach but rather stayed at the base of the mouth tube (Nicolai 1783, p. 290; Riskin 2003). The excretion process therefore was chemically unrelated to the ingested food. Nevertheless, there remain some similarities between this mechanical duck and living creatures, especially with regard to body movements.

Starting with René Descartes' mechanical understanding of life phenomena in the first half of the 17th century, the description and analysis of living beings in terms of mechanical devices was well established. In this context, the expression "artificial life" was introduced, e.g. by Thomas Hobbes in the introduction to his book "Leviathan" which first appeared in 1651. There he posed the following question:

> why may we not say, that all automata (engines that move themselves by springs and wheels as doth a watch) have an artificial life? For what is the heart, but a spring; and the nerves, but so many strings; and the joints, but so many wheels, giving motion to the whole body, such as was intended by the artificer (Hobbes 1651, p. ix).

However, obviously there are things these mechanical automata are unable to do: they do not build their own body and they do not reproduce. They are *allopoietic* instead of *autopoietic*; they depend on an external designer and constructor.

These shortcomings are crucial, of course, and as a result there were many attempts, for example by Immanuel Kant at the end of the 18th century, to specify the fundamental characteristics of living beings in contrast to automata, even if the latter show some life-like properties.

*Inorganic Chemical Processes Induced in the Laboratory as Artificial Life*
In the 19th century, progress in the physiological and chemical understanding of the phenomena of life resulted in new approaches in artificial-life research. For some leading scientists a complete understanding of life seemed to be possible in the near future. Famously, Emil DuBois-Reymond claimed, in 1872, that consciousness was too complex a problem for the human mind to solve, it belonged to the category of *ignorabimus*: we will never know. In contrast, life was, according to DuBois-Reymond, something not yet understood but in principle understandable: *ignoramus*, not *ignorabimus* (1872, p. 64–5).

Some decades later, physiologist Jacques Loeb declared that it would be possible not only to understand but to create life. In a letter to Ernst Mach in 1890 he wrote:

> The idea is now hovering before me that man himself can act as a creator, even in living nature, forming it eventually according to his will. Man can at least succeed in a technology of living substance (acc. to Pauly 1987, p. 51).

Loeb admitted that contemporary artificial machines were fundamentally distinct from living systems, but since he viewed living organisms as nothing but "chemical machines", he maintained that there is nothing that impedes "the artificial construction of living machines" (Loeb 1906, p. 1). With this claim the era of synthetic biology began, at least in its programmatic phase.

The very term "synthetic biology" is usually attributed to Stéphane Leduc, a French biologist. But in fact the expression already appeared in the middle of the 19th century. Hugh Doherty brought it into use in 1864, in the sense of "unified" or "systematic biology" (Doherty 1864, p. iv). Benjamin Franklin, the Christian writer who was the Rector of Christ Church in Shrewsbury, New Jersey, USA, used it in a similar sense in 1881 (p. 19).

But the first to use the expression in its modern sense seems to be Stéphane Leduc, in a short paper about the laws of biogenesis which appeared in 1906. He claimed that time was ripe for the new science of synthetic biology: "Nos ressources scientifiques sont actuellement suffisantes pour nous permettre, à côté de la biologie analytique, d'élever la biologie synthétique" (Leduc 1906, p. 268). Nevertheless, two years later, Leduc warned that not too much should be expected, because every science proceeds stepwise, by progressive evolution (Leduc 1908, p. 280). Synthetic biology, therefore, could not start by the construction of living beings which are in every respect similar to existing creatures. The systems Leduc discussed in his synthetic biology are simply products of inorganic chemical processes, for example crystal growth, mineral formation or osmotic effects in colloidal mixtures of salts.

Notwithstanding their simplicity these systems have, according to Leduc, something in common with living systems. They are intermediate products between the mineral and the living worlds: "des productions intermédiaires entre le règne minéral et les êtres vivants" (Leduc 1910, p. 135). Because of their intermediate position, these systems may help to overcome the isolated position of the life sciences that Leduc complained about. In fact, Leduc called biology part of the physico-chemistry of liquids (ibid., p. 6), but he did not really claim that his osmotic systems were living; they only resemble living beings. These resemblances are, according to Leduc, multiple:

(1) They have an "evolutionary existence" because they constantly change their shape; (2) they nourish themselves by a selective intake of substances from their environment; (3) they metabolize their food by changing their chemical constitution and expelling waste products; (4) they grow and develop and produce forms similar to those of real plants and fungi; and finally (5) they even have sensory capacities insofar as they react to disturbances from the environment. In principle, therefore, Leduc concludes, there is not a single attribute exclusive to living beings:

ni l'évolution, ni la nutrition, ni la sensibilité, ni la croissance et l'organisation, ni même la reproduction n'appartiennent exclusivement à la vie de façon à pouvoir la caractériser (ibid. p. 5).

It was not Leduc's primary aim to create synthetic life but to understand existent life forms by producing artificial life-like forms within physico-chemical systems: "C'est la méthode synthétique, la reproduction par les forces physiques des phénomènes biologiques, qui doit contribuer le plus à nous donner la compréhension de la vie" (Leduc 1912, p. 12; cf. Campos 2009, p. 8). This means that the synthetic method can be applied to provide a better understanding of natural living beings.

## Artificial Life as Viewed from Computer Algorithms

In this aim Leduc's project was radically distinct from the last approach to artificial life which I shall discuss here: artificial life as viewed from computer algorithms. In a well known, manifesto-like statement Christopher Langton wrote in 1989:

> Artificial Life is the study of man-made systems that exhibit behaviors characteristic of natural living systems. It complements the traditional biological sciences concerned with the *analysis* of living organisms by attempting to *synthesize* life-like behaviors within computers or other artificial media. By extending the empirical foundation upon which biology rests *beyond* the carbon-chain life that has evolved on Earth, Artificial Life can contribute to the theoretical biology by locating *life-as-we-know-it* within the larger context of *life-as-it-could-be* (Langton 1989, p. 1)

This is a purely functional approach to the definition of life. As Langton explicitly states, he views "life as a property of the organization of matter, rather than a property of the matter which is so organized" (ibid., p. 2). With this functionalism in mind, it is the aim of Langton's approach not only to *simulate*, but to *create* life in the computer, or, as he put it in 1987: "We would like to build models that are so life-like that they cease to be *models* of life and become *examples* of life themselves" (Langton 1987, p. 147).

The life-likeness of computer-based entities consists of capacities, "manifest functions" as Langton calls them (Langton 2000), that are similar to the capacities of living beings. These are, for example, movement, growth, and evolution. One famous example is the *glider* in the "Game of Life", developed by the British mathematician John Conway (cf. Gardner 1970, p. 123).

Following simple algorithmic rules this configuration of dots, a cellular automaton, can move across a computer screen. Thus, it has the life-like property of self-movement. In subsequent efforts, there were whole ecosystems created in computer worlds with much interaction between the entities. The interaction was interpreted as *competition, symbiosis, social learning, social strategies*, and so on, in biological terminology developed for the description of the interaction between natural organisms.

There were many discussions about whether cellular automata are mere *simulations* or *realizations* of living entities. Conway, at least, was convinced that these structures were really living, as he said: "On a large enough scale you would really see living configurations. Genuinely living, whatever reasonable definition you care to give to it. Evolving, reproducing, squabbling over territory. Getting cleverer and cleverer" (acc. to Levy 1992, p. 58).

But, of course, there are many people who have trouble ascribing life to entities consisting of computer software. Elliot Sober argues that software entities do not live because they do not *metabolize* in the proper sense of the word: they do not ingest matter and assimilate it into their body (Sober 1992, p. 374–5). On the other hand, others think the concept of metabolism could be modified in order to make it appropriate for software entities (cf. Lange 1996, p. 230).

One important point, that seems to be different in artificial life on the computer and natural life in our world, is the embodiment of structures. The metabolism of natural living systems results in the production of an individual body, which is a structurally coherent and functionally integrated material system. The body is constructed and maintained by selective assimilation of substances from the environment and it functions as a distinct, autonomous, that is internally produced as a level of causation, operating in addition to general physical laws and the boundary conditions generated by external agents like the software engineer.

I will not go into the details of these discussions here. The main point I want to make is that in this context, again, there surely are certain aspects of living beings that can be found in artificial life on the computer. Whether these structures merely *represent* or *realize* life remains a matter of judgment.

In summarizing these four approaches to artificial life–human art works that resemble living organisms, automata, osmotic chemical reactions, and computer algorithms–it is obvious that all of them have something in common with natural life. Nevertheless, I think everyone would hesitate to call them living.

## 4.2  Modes of Artificiality in Artificial Life Research

To give a more abstract characterization of the synthetic aspects of artificial life, I propose to differentiate between eight categories of artificiality, arranged in two broad groups, of which the first starts from living and the second from non-living precursors:

*Eight Categories of Artificiality in Artificial Life Research*

Starting from living precursors:

1. *Human beings*: Organisms with the capacity to establish a system of autonomous rules for the determination of actions that reach beyond the natural teleology of self-preservation and reproduction, for example the human being that made the Venus of Hohle Fels some 40,000 years ago.
2. *Artificially selected organisms*: Organisms in which certain traits are present because they were selected during the process of breeding.
3. *Genetically manipulated organisms*: Organisms with genetic material from other organisms (of different species) that was intentionally inserted.
4. *Organisms with artificially designed parts*: Organisms built on the basis of naturally occurring living precursors to which intentionally designed parts have been added, especially in order to prevent genetic interaction with natural organisms, for example the organisms in the xeno-layer of synthetic biology, which are cells with a chemically synthesized genome on a different basis than DNA (XNA).

Starting from non-living precursors:

5. *Artefacts with formal resemblance to living organisms*: Art works such as the Venus of Hohle Fels.
6. *Dynamic systems with capacities of living organisms*: Designed systems with some capacities typical of living beings, such as movement, metabolism, growth, or reproduction, for example 18th century automata or Leduc's osmotic vegetations.
7. *Synthetic organisms*: Intentionally constructed organized systems that have many capacities typical of living beings and which are constructed out of non-living matter. The protocell layer of synthetic biology belongs here.
8. *Nature identical synthetic organisms*: Intentionally constructed organisms that are identical, with respect to their structure and functioning, to organisms occurring in nature. Some of the future products of engineering biology would belong in this category.

In this classification, "artificiality" is a complex concept with a catalog of quite different meanings. The same is true, certainly, of "life", the second component of "artificial life", to which I shall turn later. This is, at least, one common ground for the two concepts: "artificiality" and "life" are both heterogeneous notions. But there are other similarities as well, which I shall address now.

## 4.3 Artificial Systems as Models for Living Systems

Since late antiquity, automata have been described by some authors as if they were living beings. In the fourth century AD Gregory of Nyssa gave an account of mechanical toys exhibiting phenomena typical of living beings, for example certain forms, movements, and sounds. He explains them by reference to a *moving power* (κινετικὴ δύναμις; Gregory of Nyssa, De anima et resurrectione, 33D1–36A15; Meissner 1990, p. 232–3), which was the result of a mechanism put into the system by the designer. Gregory explicitly states that, in these systems, the *mechanism* (τέχνη) replaces the soul as the moving and organizing principle of these systems (Gregory of Nyssa, De anima et resurrectione, 36B5–15).

In the 17th century, the mechanical artefact, the machine, was established as a prominent model for understanding and explaining living beings. In this process, *living beings* changed into *organisms* insofar as the concept of *organization* displaced the ancient principle of life, which is the soul. Hence, the machine was conceptually not opposed to the living being, but served as a means for the understanding of its working order. This is especially obvious in Kenelm Digby's comparison of "living creatures" and mechanical "engines". According to Digby, living creatures and engines are both heterogeneous systems consisting of different parts that are harmoniously arranged. The common organizing principle of living beings and mechanical artefacts is *interdependence* of parts:

the one [part] not being able to subsist without the other, from whom he deriveth what is needefull for him; and again being so usefull unto that other and having its action and motion so fitting and necessary for it, as without it that other can not be (Digby 1644, p. 205; cf. Cheung 2008, p. 25).

The immediate stimulus for Digby's comparison of machines and living beings was an engine for producing coins which he saw during a trip to Spain. In this machine the end product was achieved by the organized interaction of the parts; the same was true in the body of an animal, as Digby argued:

every one [part], requireth to be directed and putt on in its motion by an other; and they must all of them (though of very different natures and kinds of motion) conspire together to effect any thing that may be, for the use and service of the whole (Digby 1644, p. 208; cf. Blank 2010, p. 127ff.).

It is the comparison with man-made machines that allowed Digby to describe living creatures as harmonious systems of parts that interact, or "conspire" as he wrote, with respect to one common goal and without postulating a center of regulation.

In the 18th and 19th century the machine became a well-established model for the explanation of organisms. Famously, Immanuel Kant analyzed organisms in an analogy with man-made machines. He argued that our idea of the unity of the organism has to judge it as if it were the result of the operation of an intelligent designer.

Following Michael Polanyi's analysis in 1968, machines and organisms are both "irreducible structures" because their working order cannot be reduced to general physical laws, as it depends on the special boundary conditions which are imposed by the very structure of the system (Polanyi 1968, p. 1309–10). Machines and organisms are under "dual control", as Polanyi put it; they are controlled, as every system is, by the laws of physics and chemistry, and at the same time by their own organization which creates the boundary conditions for these laws and which constrains the dynamic processes constituting their identity (Moreno and Umerez 2000, p. 107). Hence, in machines and organisms we have a kind of downward causation, or at least downward control of the activities of the parts by the structure of the whole system. For this reason, machines and organisms together form a class of organized systems that is clearly separated from the inorganic bodies of nature. Or, as Hannah Ginsborg put it, organisms are mechanically inexplicable simply because of their machine-nature:

organisms are mechanically inexplicable, not in virtue of what distinguishes them from machines, but rather in virtue of what they have in common with machines [...]: they possess a regular structure, and display regularities in functioning, which cannot be accounted for in terms of the basic physical and chemical powers of matter alone (Ginsborg 2006, p. 462).

The reason why I discuss this conceptual vicinity of machines and organisms here is that it makes talk of "artificial" or "synthetic life" very plausible. Although living organisms are natural and machines are artificial there is, nevertheless, a structural congruence in both that has been highlighted for centuries. It is part of our traditional understanding of the concept of life that it designates systems that are structurally similar to man-made machines. This similarity refers, first of all, to the internal and external teleology of machines and organisms: both consist of functional parts and are useful for external purposes. Besides that, machines and living beings can be characterized on an organizational level by similar engineering principles, like standardization and modularization of their components. Synthetic biology, especially in its engineering layer, emphasizes this parallel. The conceptual affinities between the terms "living organism" and "machine" are so close that even the group of scientists that established *autopoiesis* as the most distinctive property of living beings in contrast to non-living things, used "living machine" as their standard, technical term to designate organisms ("máquinas vivientes"; Maturana and Varela 1972, p. 17).

Despite these conceptual affinities between living organisms and machines, synthetic biology faces a fundamental dilemma on the basis of the traditional understanding of "machine": synthetic biologists want their products to be stable and predictable in their behaviour, yet the essence of living beings is constant change and plasticity (Schark 2012, p. 32). To resolve this dilemma, the living machines of synthetic biology may themselves become plastic and autonomous, at least to a certain degree. By this development, synthetic biology will not only change our view of the phenomenon of life but also our concept of machines: with their status of living beings, the products of synthetic biology will have an autonomy and a value of their own which exists alongside our instrumental interests that we have by using them. Linked to this, the artificial living beings will have interests of their own which may demand that we have duties towards them. In conclusion, with the creation of living machines synthetic biology will not necessarily strengthen our purely mechanistic view of life but rather broaden our concept of machines.

## 4.4 Modes of Vitality in Artificial Life Research

One of the most characteristic features of biology is that some of its central concepts have a very long tradition, reaching all the way back to antiquity. In contrast to physics, biology has not faced conceptual revolutions in which the core of its central concepts was overthrown. 'Nutrition', 'growth', 'development', 'reproduction', or 'sensation' are all concepts that were already present in ancient theories of life. And even the one theory that is often associated with a revolution in biology, the evolutionary theory of the 19th century, has not changed these concepts much. Many physiological theories and concepts remained largely untouched by this so-called "revolution" in biology. One of the reasons for this may be that many fundamental concepts in biology are *descriptive, functional concepts*: they conceptualize activities of organisms by their visible effects or functions. Even the detailed physiological explanations of mechanisms, which were provided by subsequent theories, did not change much the descriptive content of these concepts:

'Nutrition', 'growth', 'development', 'reproduction', or 'sensation' are fundamental biological phenomena that were conceptualized before they could be explained, and their explanations did not significantly alter their descriptive content. All this is also true of the most fundamental concept of biology, which is, of course, "life".

"Life" is a descriptive concept that was introduced in antiquity and was defined on the basis of certain capacities and functions of organisms. In a famous passage in *De anima*, which could be considered as the starting point of the scientific use of the concept, Aristotle gives the following explanation:

> Of natural bodies some have life and some have not; by life we mean the capacity for self-nutrition, growth and, decay. [...] The word living is used in many senses, and we say that a thing lives if any one of the following is present in it – mind, sensation, movement or rest in space, besides the movement implied in nutrition and decay or growth (Aristotle, On the soul 412a; 413a.).

In this passage, the concept is introduced not by identifying one single property but by citing several. These properties do not always occur together. Aristotle himself admits that plants do not have sensations and do not move. That is the reason why he maintains that they have a different kind of life compared to animals: "life" is generally understood to mean not one kind of thing only, but to be one thing in animals and another in plants" (Aristotle, Topics 148a30–31). Consequently, the word "life" is, and Aristotle explicitly says this, homonymous.

Since antiquity the homonomy of the term "life" still has continued to increase. For Aristotle it was clear that life was a property or, more precisely, a way of being of an individual organism. In subsequent times, the focus shifted away from the individual to the series of organisms in time and the concept "life" was increasingly associated with this progression in time. This trend was especially strong after the formulation of the theory of evolution. For the neo-Darwinian biologist August Weismann it is not the individual organism that has a *proper life* ("eigentliches Leben"), but the germ cells that have a potentially indefinite existence. Compared to that, Weismann calls the individual body of an organism a *secondary appendage* ("nebensächliches Anhängsel"; Weismann 1884, p. 165). A similar view can be found in Paul Valéry when he writes: "Life changes one individual for another as a moving object changes surroundings, as a traveller changes trains" (Valéry 1923, p. 140; this simile has ancient roots: Lucretius said of life it is passed on from one body to the next like a torch: "quasi cursores vitai lampada tradunt"; De rerum natura II, 79). The individual organism is only "half of life", as Jesper Hoffmeyer and Claus Emmeche wrote in 1991 (p. 154). The other half is a process embracing more than individual organisms.

This reference to individual organisms as well as interindividual processes makes the concept of life fundamentally ambiguous. Already in ontological terms it is not at all clear what the referent of "life" is. Here is a short list of ontological categories to which the referent of "life" may belong (Toepfer 2011, tab. 158):

1. The ontologically irreducible way of being of individual organisms.
2. The sum of characteristic types of activities of organisms, e.g. nutrition, growth, reproduction.

3. The arrangement of matter or material state of a certain class of objects.
4. A property of complex dynamic systems far from equilibrium.
5. The sequence of events and their unity in the life-history of an individual organism.
6. The sum of living beings in a geographical region or space of time.
7. A multigenerational dynamic in the "chain" or "flow" of beings in time.
8. A measure for the amount or intensity of the characteristic activities of organisms.

With all these different meanings in mind, it is difficult to make use of the word in order to achieve a clear understanding of a phenomenon. Hence, "life" may be a useful concept for integrative purposes because it incorporates the many facets of biological phenomena, but it is not very useful for clear argumentations in specific contexts. It has the vague character of an overarching umbrella-concept, not of a scientific term (cf. Machery 2012).

The inherent heterogeneous nature of the concept of life was obvious already at its historical beginning in Aristotle. This heterogeneity has increased since then, especially with the additional perspective of evolutionary theory. Therefore, one may say that scientific progress has resulted in a progressive dissolution of the concept of life. The word does not designate any clearly defined object.

Synthetic biology may have a strengthening influence on some of the aspects of this concept while suppressing others, because its primary objects of concern are individual living beings. The approach of synthetic biology is compatible with the thesis of the irreducibility of vital phenomena to the level of the system's components; in both natural and artificial organisms life consists of an ensemble of activities that is only present at a high level of integration in an individual system. Nevertheless, synthetic biology is, in all its different layers, especially focused on the compositional aspects of organisms. It deals with molecules, compartments and local mechanisms in order to create the self-maintaining bodies of living beings and their characteristic types of activities. With this organismic and individualistic approach, the sequence of organisms in time and the possibility of evolution is rather neglected, or, for reasons of safety and sustainable use, even intentionally suppressed. Life in synthetic biology is the reproducible set of functions and activities resulting from the composition of individual organisms that ideally remains the same across generations.

## 4.5 Can Artificially Designed Systems Be Instantiations of Life?

In the third section of this chapter I pointed out the conceptual proximity between "life" and "machine". This is, certainly, only part of the story. For many contemporary authors "artificial life" is an oxymoron, a contradiction in terms, the one belonging to nature, the other to human culture (Mutschler 1999–2000, p. 73;

Rödl 2014). Because artificial life is intentionally planned and constructed, it is distinct from the spontaneous origin of natural living organisms. Whereas the machine is the paradigm of the completely controlled system, the entirely understood object, the living organism remains an enigma in its self-referential organization, spontaneous behavior and subjective experience of its particular world.

Nevertheless, as I pointed out in section three, there are close conceptual affinities between "life" and "machine". Both are functional concepts, and usually they are identified by types of activities and the outcome of an inner working order. They do not depend on a certain type of matter or a particular causal history (natural or artificial). It is only basic functions that identify a system as living. Therefore, in contrast to "nature" which always had its counterpart in "technology" or "culture", "life" is, and since its beginning has always been, a concept that, at least in our imagination, connects the two sides of the nature-culture divide. For conceptual reasons there never will be artificial nature, but since antiquity there has always been the conceptual possibility of artificial life.

Besides this fundamental conceptual level, the artefact model had always had immense heuristic value for research in biology. Even many fundamental theoretical concepts of biology can best be analyzed within the artefact model, the *technomorph* understanding of living organisms, as Ernst Topitsch called it (1958, p. 19). Subsequently, philosophers of biology were of the opinion that any theoretical explication of central biological concepts, such as 'function', depend on the technomorph perspective (Ruse 1982, p. 304). This "engineering perspective on biology" was thought to be fundamental also for analyzing the transformations that have taken place in the evolutionary past of all organisms (Dennett 1995, p. 233).

But there is, of course, one ironic twist in all these statements: in the last 150 years the evolutionary perspective has gained an increasing influence on biological thinking. This perspective was included in the very definition of the concept of life (Muller 1966, p. 513; Joyce 1994, p. xi; Ruiz-Mirazo, Peretó and Moreno 2004, p. 330). In these definitions, evolution comes into play as future evolution; it is not claimed that past evolution is essential. But there are other authors who claim this as well (Lahav and Nir 2002, p. 133; Dupré and O'Malley 2009, p. 1).

Nevertheless, there are many definitions of "life" with no reference at all to evolution (Szathmáry 2002, p. 183; Damiano and Luisi 2010, p. 149). The irony with past evolution as a criterion for life is that the genealogical connection of past evolution is the only sufficient criterion for the identification of a system as living. Being part of the phylogenetic tree is the only exclusive property common to all currently existing forms of life on earth. In contrast to this property of genealogical association, the other signs of life, self-motion, growth, or reproduction, are not exclusive properties of living beings. This was already declared by Stéphane Leduc, whom I quoted at the beginning. But, and this is the irony, being part of one and the same phylogenetic tree may be a sufficient but not a necessary condition for a system to be alive. This was evident in biology before the advent of evolutionary theory, of course, as there were supposed to be manifold instances of spontaneously generated organisms, but it was well established even after the general acceptance of evolutionary theory, as the existence of the science of exobiology, that is of hypothetical extraterrestrial life, proves.

It seems that man will turn this science of living beings, that are not part of the genealogical tree of life on earth, from its hypothetical stage into a science dealing with real entities. By doing so, he will remove the only criterion that is common to all forms of life on earth at the moment, which is common descent. After the creation of the first synthetic organism, the universal genealogical connection of living beings will be the restricted empirical criterion only for life as it can be found in nature (natural life), and not for life per se.

# References

Anonymous (1868) Physiognomy Anthropol Rev 6: 137–154
Anonymous (1870) Review: The theory of the arts. Anthropol Rev 8: 145–162
Aristotle (1928) Topics. In: Topica and De sophisticis Elenchis; transl. by WA Pickard-Cambridge. Clarendon Press, Oxford
Aristotle (1936/1957) On the soul; transl. by WS Hett. Harvard University Press, Cambridge, Mass
Blank A (2010) Biomedical Ontology and the Metaphysics of Composite Substances 1540–1670. Philosophia, Munich
Boldt J, Müller O, Maio G (2009) Synthetische Biologie. Eine ethisch-philosophische Analyse, Bundesamt für Bauten und Logistik, Bern
Brenner A (2012) Leben leben und Leben machen. Die Synthetische Biologie als Herausforderung für die Frage nach dem Lebensbegriff. In: Boldt J, Müller O, Maio G (eds) Leben schaffen? Philosophische und ethische Reflexionen zur Synthetischen Biologie. Mentis, Paderborn, pp 105–120
Campos L (2009) That was the synthetic biology that was. In: Schmidt M et al (eds) Synthetic Biology. The Technoscience and Its Societal Consequences. Springer, Dordrecht, pp 5–21
Caplan A (2010) The end of vitalism. Nature 465: 423
Chapuis A, Droz E (1958) Automata. A Historical and Technological Study. Griffon, Neuchatel
Cheung T (2008) Res vivens. Agentenmodelle organischer Ordnung 1600–1800. Rombach, Freiburg i. Br
Conard NJ (2009) A female figurine from the basal Aurignacian of Hohle Fels Cave in southwestern Germany. Nature 459: 248–252
Damiano L, Luisi PL (2010) Towards an autopoietic redefinition of life. Orig Life Evol Biosp 40: 145–149
Dennett DC (1995) Darwin's Dangerous Idea. Evolution and the Meaning of Life. Penguin, London
Deplazes-Zemp A (2012) The conception of life in synthetic biology. Sci. Eng. Ethics 18: pp 757–774
Digby K (1644) Two Treatises in the one of which the Nature of Bodies, in the other, the Nature of Man's Soule is Looked into. Blaizot, Paris
Doherty H (1864) Organic Philosophy or Man's True Place in Nature, vol 1. Epicosmology. Trübner, London
DuBois-Reymond E (1872) Über die Grenzen des Naturerkennen. In: Vorträge über Philosophie und Gesellschaft. Meiner, Hamburg 1974, pp 54–77
Dupré J, O'Malley MA (2009) Varieties of living things: life at the intersection of lineage and metabolism. Philos Theory in Biol 1: 1–25
Franklin B (1881) The Creed and Modern Thought. Young, New York

Gardner M (1970) The fantastic combination of John Conway's new solitaire game "life". Sci
    Am 223 (4): 120–123
Ginsborg H (2006) Kant's biological telelology and its philosophical significance. In: Bird G (ed)
    A Companion to Kant. Blackwell, Malden, Mass, pp 455–469
Goethe JW (1795) Erster Entwurf einer allgemeinen Einleitung in die vergleichende Anatomie,
    ausgehend von der Osteologie. In: Kuhn D (ed) (1954) Goethe. Die Schriften zur
    Naturwissenschaft. Deutsche Akademie der Naturforscher Leopoldina, vol. I, 9. Morphologische
    Hefte. Böhlau, Weimar, pp 119–151
Gregory of Nyssa (c. 380 AD) De anima et resurrectione. In: Patrologia Graeca, vol. 46, ed. by
    JP Migne. Paris 1863, coll. 12–160
Heckmann H (1982) Die andere Schöpfung. Geschichte der frühen Automaten in Wirklichkeit
    und Dichtung. Umschau, Frankfurt/M
Hegel GWF (1820/29) Vorlesungen über die Ästhetik, vol. 1. In: Werke, vol. 13. Suhrkamp,
    Frankfurt/M 1970
Hobbes T (1651) Leviathan or the Matter, Forme and Power of a Commonwealth Ecclesiastical
    and Civil. In: The English Works, vol. 3. Bohn, London 1839
Hoffmeyer J, Emmeche C (1991) Code-duality and the semiotics of nature. In: Anderson M,
    Merrell F (eds) On Semiotic Modeling. Mouton de Gruyter, Berlin, pp 117–166
Joyce G (1994) Foreword. In: Deamer DW, Fleischaker GR (eds) Origins of Life. The Central
    Concepts. Jones and Bartlett, Boston, pp xi–xii
Lahav N, Nir S (2002) Life's definition. In search for the most fundamental common denomi-
    nators between all living entities through the entire history of life. In: Palyi G, Zucchi C,
    Caglioti L (eds) Fundamentals of Life. Elsevier, Paris, pp 131–133
Lange M (1996) Life, "artificial life", and scientific explanation. Philos Sci 63: 225–244
Langton CG (1987) Studying artificial life with cellular automata. Physica 22D:120–149
Langton CG (1989) Artificial life. In: Life Artificial (ed) id. Addison-Wesley, Redwood City,
    Calif., pp 1–47
Langton CG (2000) A new definition of artificial life. http://www.chairetmetal.com/cm03/intro.htm
Leduc S (1906) Les lois de la biogénèse. La Revue scientifique 5 (5. sér.): 265–268
Leduc S (1908) Essais de biologie synthétique. In: Festband der Biochemischen Zeitschrift für
    H.J. Hamburger, pp 280–286
Leduc S (1910) Théorie physico-chimique de la vie et générations spontanées. Poinat, Paris
Leduc S (1912) La biologie synthéthique. Poinat, Paris
Levy S (1992) Artificial Life. A Report from the Frontier where Computers Meet Biology.
    Vintage Books, New York
Loeb J (1906) Vorlesungen über die Dynamik der Lebenserscheinungen. Barth, Leipzig
Machery E (2012) Why I stopped worrying about the definition of life… and why you should as
    well. Synthese 185:145–164
Maturana HR, Varela FJ (1972) De máquinas y seres vivos. Una teoría sobre la organización
    biológica. Santiago de Chile; Engl.: Autopoiesis and Cognition. The Realization of the
    Living. Reidel, Dordrecht 1980 = Boston Studies in the Philosophy of Science, vol. 42
Meissner HM (1990) Rhetorik und Theologie. Lang, Frankfurt/M
Moreno A, Umerez J (2000) Downward causation as the core of living organization. In: Andersen
    PB et al. (eds) Downward Causation. Minds, Bodies and Matter. Aarhus University Press,
    Aarhus, pp 99–117
Muller HJ (1966) The gene material as the initiator and the organizing basis of life. Am Nat 100:
    493–517
Mutschler H-D (1999–2000) Die Technisierung des Lebendigen. Über „Künstliches Leben".
    Scheidewege 29: 72–102
Nicolai F (1783) Beschreibung einer Reise durch Deutschland und die Schweiz im Jahre 1781,
    vol. 1. Selbstverlag, Berlin
Pauly PJ (1987) Controlling Life–Jacques Loeb and the Engineering Ideal in Biology. Oxford
    University Press, New York

Petrus Damiani (2002) Epistula 44. In: Opere, vol. I, 3, ed. by GI Gargano. Città Nuova, Rome
Plessner H (1928) Die Stufen des Organischen und der Mensch. de Gruyter, Berlin 1975
Polanyi M (1968) Life's irreducible structure. Science 160: 1308–1312
Rheinberger HJ, Bredekamp H (2012) Die neue Dimension des Unheimlichen. In: Köchy K,
    Hümpel A (eds) Synthetische Biologie. Entwicklung einer neuen Ingenieurbiologie? Berlin-
    Brandenburgische Akademie der Wissenschaften, Berlin, pp 162–163
Riskin J (2003) The defecating duck, or, the ambiguous origins of artificial life. Critical Inquiry
    29: 599–633
Rödl S (2014) Leben herstellen. Deutsche Zeitschrift für Philosophie 62: 74–89
Ruiz-Mirazo K, Peretó J, Moreno A (2004) A universal definition of life: autonomy and
    openended evolution. Origins of Life and Evolution of Biospheres 34: 323–346
Ruse M (1982) Teleology redux. Boston Studies in the Philosophy of Science 67: 299–309
Schark M (2012) Synthetic biology and the distinction between organisms and machines.
    Environmental Values 21: 19–41
Schummer J (2011) Das Gotteshandwerk. Die künstliche Herstellung von Leben im Labor,
    Suhrkamp, Frankfurt/M
Sober E (1992) Learning from functionalism–prospects for strong artificial life. In: Boden MA
    (ed) (1996) The Philosophy of Artificial Life. Oxford University Press, Oxford, pp 361–378
Szathmáry E (2002) Units of evolution and units of life. In: Palyi G, Zucchi C, Caglioti L (eds)
    Fundamentals of Life. Elsevier, Paris, pp 181–195
Toepfer G (2011) Leben. In: Historisches Wörterbuch der Biologie. Geschichte und Theorie der
    biologischen Grundbegriffe, vol. 2. Metzler, Stuttgart, pp 420–483
Topitsch E (1958) Vom Ursprung und Ende der Metaphysik. Eine Studie zur
    Weltanschauungskritik. Springer, Wien
Valéry P (1923) Fragment. In: Stimpson B (ed) Cahiers = Notebooks, vol. 4. Lang, Frankfurt/
    M 2010
Valéry P (1910) Cahier B 1910. Gallimard, Paris 1926
Vaucanson J (1738) Le mécanisme du fluteur automate, avec la description d'un canard artificiel.
    Guerin, Paris
Von Hufeland CW (1796/1798) Die Kunst das menschliche Leben zu verlängern, vol. 2.
    Akademische Buchhandlung, Jena
Weismann A (1884) Über Leben und Tod. In: Aufsätze über Vererbung und verwandte biologische
    Fragen. Fischer, Jena 1892, pp 123–190

# Chapter 5
# New Debates in Old Ethical Skins

Christian Illies

> *I would feel more optimistic about a bright future for man*
> *if he spent less time proving that he can outwit Nature*
> *and more time tasting her sweetness and respecting her*
> *seniority.*
> E.B. White (2007)

**Abstract** One of the central characteristics of modern science is that understanding can be measured by prediction, manipulation, and manufacture. That is why synthetic biology can be regarded as the climax of this project. It would seem that we truly understand living organisms only when we can re-create them. Synthetic biology therefore radicalizes tendencies in modern science. What does this mean for ethics? By applying five criteria, originally proposed by Hans Jonas, we can show that synthetic biology is of high ethical relevance. To illustrate this relevance, traditional ethical issues, such as risk and outcome-uncertainty, are applied. The 'playing God' argument is scrutinized by dividing it into three sub-arguments: the 'hubris' argument (science overestimates its power), the 'taboo' argument (some things should not be done at all), and the 'sorcerer's apprentice' argument (we might not be able to control what we create). None of these concerns are entirely new; they are known, for example, in genetic engineering. But they seem to reach new significance in the realm of synthetic biology. As a consequence, precautionary principles are most certainly demanded by synthetic biology and a new ethical debate is called for. This debate is urgent but also particularly challenging. It raises fundamental issues (such as the meaning of basic terms like 'nature') and faces professional, conceptual, and institutional inhibitions which must be overcome for substantial answers to be reached. The chapter concludes by suggesting three ethical laws which might be taken as basic for any future ethics of synthetic biology.

---

C. Illies (✉)
Chair of Philosophy II, Otto Friedrich University, 96045 Bamberg, Germany
e-mail: christian.illies@uni-bamberg.de

© Springer International Publishing Switzerland 2016
M. Engelhard (ed.), *Synthetic Biology Analysed*, Ethics of Science
and Technology Assessment 44, DOI 10.1007/978-3-319-25145-5_5

1.: A biotic artefact may not injure a human being. 2.: A biotic artefact must be strictly functional, except where this would conflict with 1. And 3.: A biotic artefact must be protected and respected as a form of life, so long as such protection does not conflict with either 1 or 2.

## 5.1 Ethics of Technology

Most ethical debates begin only when events urgently require an ethical or political answer.[1] Novel cultural phenomena, including new technologies and scientific discoveries, often pose problems for ethics that were not obvious at the outset. A long time passed, for example, before the effects of industrialisation were generally regarded as a global threat to the climate.[2] The negative effects of high $CO_2$ concentration in the atmosphere were already pointed out in 1980 when *Global 2000*, the report for the American president, was published. The report clearly speaks about carbon dioxide's "potential for warming the earth" (Seven Locks edition; p. 36). But initially no *ethical* discussion followed; for the first three decades climate change remained a topic of economic and scientific debate only (cf. Brown 2012).

It is hard to deny that technological innovations require ethical reflection. Few things have changed our world more fundamentally than science and technology; they alter the face of the non-human world, but they also re-shape human activities and their meaning (cf. Mitcham and Waelbers 2009). Winner (1980) argues that artefacts "have politics" because they change social arrangements and re-order the social world. And, most obviously, nothing magnifies the effects of human action more dramatically than technology. Traditional ethical theories have almost exclusively focused on society, human well-being, and human interaction on a short-range scale (both in space and time) and were not designed for these new types of problems: "Modern technology has introduced actions of such novel scale, objects, and consequences that the framework of former ethics can no longer contain them", as Hans Jonas points out (Jonas 1984, p. 6).

It seems a truism that phenomena can be addressed and analysed only after they have become apparent. We have to know what it is we are talking about. That is why Hegel asserts that philosophy understands its own time only with hindsight.[3] But moral orientation should accompany and guide developments rather than write their obituaries. This is obviously a demanding task, requiring a wide competence. Besides having a normative understanding, ethicists must identify consequences of technologies in advance, by, for example, learning from experience gained in

---

[1]The Vietnam War, for example, triggered an intense debate on the use of power, just war, and just society. Rawls' *Theory of Justice* (1971) can be read as a response to this event.

[2]A general awareness of environmental impact began earlier, with, for example, Rachel' Carson's *Silent Spring* (1962) and the studies of the *Club of Rome*.

[3]This is captured by Hegel's melancholic remark in the Preface of his *Philosophy of Right* (1820): "the owl of Minerva spreads its wings only with the falling of the dusk".

familiar cases (familiarity principle). Such work can hardly be done by ethicists alone and, luckily, there are many interdisciplinary engagements where the ethical problems of new technologies are tackled (such as the research conducted in the course of writing this book). That moral orientation requires profound factual knowledge has been widely acknowledged and has given rise to, for example, the new science of Risk Analysis (RA) and of Technology Assessment (TA) in the 1970s (Hansson 2009). By then it had become obvious that many technological developments, in particular the large-scale use of industrial chemicals such as pesticides and fertilizers, have harmful effects on the environment in the long-term.[4]

It is obvious that synthetic biology could also have harmful effects; it has the potential to change things radically. It is therefore desirable that ethical reflection accompanies synthetic biology from the outset. And luckily synthetic biology been accompanied by ethical reflection from the very beginning.

## 5.2  Synthetic Biology—A Philosophical View

### 5.2.1  Synthetic Biology as a Climax of Modern Science and Technology

In many ways, synthetic biology is more than just an example of a new technology in need of ethical orientation—it is a *climax* of modern science and technology. Weber (1934) and White (1978) point out that new cultural ideas and ideals were important impulses for the Scientific Revolution in 16th and 17th century Europe. Following Vittorio Hösle (1991, pp. 43–68), we might identify three central ideas:

(a)  **New self-understanding**
   Mankind began to consider itself in a new way in the 15th century (cf. Burckhardt 1878). Renaissance humanists and theologians understood human beings to have a unique status within nature due to their freedom, reason, and creative powers (cf. Na 2002, p. 93). This idea culminated in the Renaissance ideal of art and the artist.

(b)  **Changed conception of nature**
   The more human beings saw themselves as self-determining, independent subjects, the more nature appeared as a well ordered system of events without any intent or *telos*. Simmel (1900, p. 30ff.) observes that the main difference between

---

[4]At the same time, the general acceptance of modern technology declined and mistrust and disapproval (for example, of nuclear power) became quite common in the West (e.g. Habermas, Marcuse, Schelsky, but also Pardo and Hagen in this volume): "Without this crisis [of orientation] surrounding the optimistic belief in progress, TA would presumably never have developed" (Grunwald 2009, p. 1105). The pressure became too high and politics realized that it had to find answers—which required a more detailed knowledge and understanding of the societal and environmental impacts of concrete technical products and processes (TA), and of the probability and severity of possible damage (RA).

antiquity and modernity is the "deep and profound notion of the self" that is opposed to the "strong notion of objectivity" in nature "that finds its expression in the non-circumventable laws of nature". These laws were seen as "mechanical", representing a causal and non-intentional order that can be expressed in mathematical language.[5]

(c) **Understanding of truth**

A new concept of truth resulted from this altered self-understanding. The traditional concept relied upon the existence of things independent of (but discovered by) rational beings; the new concept emphasized dependence on human beings. Something is 'true' if it is at our disposition, if we can master it. Giambattista Vico expresses this as the Verum-factum Principle first formulated in 1710 as part of his *De antiquissima Italorum sapientia, ex linguae latinae originibus eruenda*: "The criterion and rule of the true is to have made it. Accordingly, our clear and distinct idea of the mind cannot be a criterion of the mind itself, still less of other truths. For while the mind perceives itself, it does not make itself." The point is that we can only understand something that we can also recreate (Löwith 1968 and Na 2002)—and thus truth is verified through bringing something about, not by mere observation or clear insights (as Descartes had argued). The true and the made are convertible.

This point has been taken up by modern theory of science. The so-called Verification and Falsification Principles (of Logical Positivism) follow this line but shift emphasis to us as human agents, since our *actions* determine whether or not we can call some hypothesis "true" (Löwith 1986). Statements about cause and effect are only considered to be correct if we can replicate the phenomenon they refer to, that is, if *human beings can bring the effect about in an experiment*.[6] Or as Richard Feynman, the late Nobel Laureate in physics, had written on his blackboard at the time of his death in 1988: "What I cannot create, I do not understand". (Craig Venter's team famously coded these words, though slightly misquoted, into their first chemically synthetized bacterial genome.[7]) In sum, a proper science is based upon a kind of technical knowledge, a knowledge about how to change or

---

[5]The "book" of the universe "is written in the mathematical language" as Galilei remarked in 1623 (1842, p.171). Quoted from Donald DeMarco (1986).

[6]It is not a sufficient condition because we can sometimes make things which we do not understand.

[7]Cf. Press release from the J. Craig Ventre Institute (20th May 2010): "they designed and inserted into the genome what they called watermarks. These are specifically designed segments of DNA that use the "alphabet" of genes and proteins that enable the researcher to spell out words and phrases. [...] Encoded in the watermarks is a new DNA code for writing words, sentences and numbers. In addition to the new code there is a web address to send emails to if you can successfully decode the new code, the names of 46 authors and other key contributors and three quotations: 'TO LIVE, TO ERR, TO FALL, TO TRIUMPH, TO RECREATE LIFE OUT OF LIFE.'—JAMES JOYCE; 'SEE THINGS NOT AS THEY ARE, BUT AS THEY MIGHT BE.'—a quote from the book, 'American Prometheus'; 'WHAT I CANNOT BUILD (sic!), I CANNOT UNDERSTAND.'—RICHARD FEYNMAN." See: http://www.jcvi.org/cms/press/press-releases/full-text/article/first-self-replicating-synthetic-bacterial-cell-constructed-by-j-craig-venter-institute-researcher/#sthash.DPPZiRxd.dpuf.

how to make things.[8] This explains the close connection of modern science and technology: the process of scientific discovery is itself a technical one. Francis Bacon, another father of the modern scientific project, states this point rather plainly when he claims that knowledge is power (*"ipsa scientia potestas est"*).[9] By this he means that knowledge is not about understanding the world as it is but about changing it for the better. (Or, as we might add, sometimes for the worse.)

The structure of modern science clarifies why physics and chemistry could much more easily (than biology) become modern sciences and explains the basis for their respective technologies (Illies 2006a, b, p. 50 ff. See also Chap. 2 of this volume.). It is relatively unproblematic to design and execute physical or chemical experiments, while life phenomena are much more difficult to manipulate. Kant expresses the problem in his *Critique of Judgement*, where he doubts that organisms would ever give up their secrets: "it is absurd for human beings [...] to hope that perhaps some day another Newton might arise who would explain to us, in terms of natural laws unordered by any intention, how even a mere blade of grass is produced." (Kant 1790 (Akademie Ausgabe Vol V, p. 400), cf. Boldt and Müller 2008, pp. 387–389).

It was only when Alfred Russel Wallace and Charles Darwin presented their theory of evolution through natural selection that biology began to become a modern science. The novelty of their "one long argument", as Darwin calls it (1993, p. 140), was not the claim that evolution has taken place (many had argued so before them), but the suggestion of a *causal* mechanism for it, namely Natural Selection. For the first time, evolutionary theory was a scientific theory in the modern sense. It promised to explain the process of evolution 'in terms of natural laws unordered by any intention'. At least it came close to explaining it because evolutionary theory could still not test its central hypothesis by controlled experiments. A proper experiment in evolutionary biology would take thousands or millions of years (at least with multicellular organisms[10]). To this extent, the project of biology as a modern science was not completed. Certainly, a crucial step in this direction has been made possible by genetic engineering and DNA-synthesis: life (that is living organisms) could be technically changed.[11] Now human beings could in principle bring about new evolutionary developments—hand-tailored variations could be experimentally "made" and thus radically altered life-forms created. One might, however, object that traditional genetic engineering merely *modifies* existing organisms; it does not create life.

By going beyond this last limitation, synthetic biology promises to be the next and decisive step towards the convergence of knowing and making. Synthetic

---

[8]This is not true for all sciences. Astronomy, for example, has long been considered to be a science without being able to perform any experiments. Here, the predictive power of theories replaces the need to bring about effects.

[9]*Meditationes sacrae* 1597, 11th section, "De Haeresibus", quoted from Bacon 1711, p. 402.

[10]Experiments with bacteria such as E. coli can take less time. It is therefore mainly with bacteria that *evolutionary experiments* are performed.

[11]One might object that domestication and breeding are already a "craft-based technology" that "was used in transforming living organisms to become biotic artefacts" (Lee 2009, p. 101). But genetic engineering is much more efficient and systematic, and allows for conscious intervention into life-forms.

biology has reached a new depth of intervention, as we showed in Chap. 2. "The deeper the science, the deeper becomes the level of manipulation through its corresponding technology, and therefore also the deeper the level of artificiality embodied by its products" (Lee 2009, p. 102; see also Toepfer in this volume). For the first time biology aspires actually to *make life from scratch* (Boldt et al. 2009, p. 48). This high aspiration looks different when we consider the three distinct approaches within synthetic biology that we have distinguished above. With respect to proto-cell research the focus is on constructing simple life forms from non-living matter. Here synthetic biology is, in a way, imitating prebiotic evolution. (As we have seen in the discussion of different definitions of life, the point at which a proto-cell can be called a living entity is certainly controversial.) With respect to engineering biology, the act of creation is mainly an act of composition; biobricks are added to a minimal cell. At this level, synthetic biology imitates existing life-forms. By contrast, with respect to chemical synthetic biology (orthogonal life or xeno-life), synthetic biology creates the very code of life anew (namely non-DNA or non-RNA xeno-nucleotides as information-storing biopolymers). If xeno-life organisms ever become alive, then biology will have transgressed the "genealogical tree of life" in a radical way (see Toepfer, Sect. 4.5.).

Even if these three approaches are different ways of "making" life, they all bring crucial phenomena for the first time under our control. Therefore the creation of new life-forms can rightly be called "one of the most important scientific achievements in the history of mankind" (Caplan 2010, p. 423). Synthetic biology promises to be the climax of modern science and technology.

### 5.2.2 The Ethical Relevance of Synthetic Biology

Synthetic biology has a prime place within the sciences and technology—but is it also of particular ethical relevance? It is helpful to look at five aspects that Hans Jonas identifies (1987); these are said to explain the challenge modern technology presents for ethics.

1. The consequences of modern technology are inherently ambivalent (1987, p. 42f.), not simply because they can be used for good or ill, but more profoundly. Even the apparently good use of technology is likely to have negative consequences in the long run.[12] Chemical fertilizers, for example, though used for good purpose, have been shown to destroy the soil by promoting erosion.
2. There is an "enforced" application of new technologies (Jonas ibid., p. 44). Once something is technically possible its application seems inevitable.

---

[12]This is a one-sided ambivalence; he does not explore the possibility that bad intent might ultimately have good consequences.

This fascination with doing what is technically possible seems to be a psychological trait in human beings. It is reflected in Robert Oppenheimer's famous statement about the atomic bomb. "When you see something that is technically sweet, you go ahead and do it and argue about what to do about it only after you've had your technical success."[13]

3. The consequences of technology seem to be inherently global and long-lasting (p. 45ff.). New technologies can create changes globally and for the long term. Climate change is an example.

4. Jonas diagnoses that technology poses problems that cannot be answered by traditional (anthropocentric) ethics. The effects of technology bring to the fore our greater responsibility: for the environment and to non-human organisms.

5. Modern technology has metaphysical relevance. In the face of the global threat to the environment Jonas asks, for example, whether and why there should be human life at all. Is there something intrinsically "good" about mankind? A positive answer cannot simply be based upon the preferences of human beings; it requires more principled (and thus metaphysical) reasoning.[14]

What Jonas claims with respect to technology in general is true for synthetic biology in particular. The inherent ambivalence of technology's consequences (Jonas 1987, p. 42f.) is certainly to be found in synthetic biology. If synthetic microbes produce renewable liquid fuels at low prices we might be able to become more independent of fossil energy by using biomass. This could, however, have unwanted consequences, such as a further increase in car production, an expansion of the private-transport infrastructure, and even more sprawling cities. There is also the unpredictability of synthetic life itself. Would all synthetic organism remain for ever under human control? Synthetic organisms can replicate and can have profound effects on the environment (see Chap. 3). Even xeno-life, separated from all other life-forms by a genetic firewall (Marliere 2009), could become uncontrollable by being practically invulnerable to any infection or disease once escaped from the laboratory. And even if the firewall prevents any gene-flow into wild populations, in principle it could numerically outgrow and replace traditional organisms by competing for resources or by releasing toxic and un-degradable compounds. The long-term effects of synthetic biology might be much worse than those of any known invasive species.

Jonas' second observation is that new technologies have their own dynamics (1987, p. 44). Obviously, synthetic biology is also "technically sweet" and seductive. One has only to look at the International Genetically Engineered Machine (iGEM) student competition that has grown, within nine years, from a local project to a worldwide synthetic biology competition. It is no longer restricted to undergraduates at the Massachusetts Institute of Technology, but includes

---

[13]Testifying in his defense in his 1954 security hearings (p. 81 of the official transcript) Quoted from http://en.wikiquote.org/wiki/Robert_Oppenheimer (20.7.2013).

[14]For general metaphysical questions raised by technology see also Dupuy (2009, p. 214).

divisions for high school students and software teams from all continents. Synthetic biology is not only "sweet", but also an intellectual challenge for many people who are not professional scientists. It is something that can be done in a backyard biological laboratory with a small budget—and it is already practiced there on a basic level.

In addition, Jonas argues that technology can readily have global and long-lasting consequences (1987, p. 45ff). This is certainly the case with synthetic life-forms; it is even the reason for creating them. Many synthetic biology companies hope to produce, by means of synthetic biology, engineered micro-organisms and specialized enzymatic products for a broad and long-lasting application. Their aim is to operate on the large scale: to produce, for example, bio-fuels using under-utilized desert land with ready nearby sources of $CO_2$ (cf. Connor and Atsumi 2010). If such technology were to succeed at low cost, then we might very quickly see a global phenomenon that is likely to become more significant in the future. If only some of the aspirations of synthetic biology are fulfilled, the face of this planet might be changed in profound ways.[15] As living beings, new organisms pose an enormous risk: It might be very difficult, if not impossible, to prevent biotic artefacts having harmful consequences.

Synthetic biology will not only transcend anthropocentric ethics (Jonas' fourth point), but transcend also traditional environmentalist ethics of various types. In recent years, there have been many debates on the question as to whether a living being has integrity, is 'intrinsically' valuable (in the way that a non-living being is not). Different positions are struck in this debate (cf. report EKAH 2010, pp. 15–19). Some critique the concept of intrinsic value as speculative or ill-founded, while others go so far as to attribute it even to micro-organisms "on the basis of conation, along with their enormous instrumental value" (Cockell 2005, p. 375). With synthetic biology, this debate is stretched even further because we are obliged to consider the value of biotic artefacts (cf. Heaf and Wirz 2002). Currently we create only micro-organisms which might be of little practical relevance, but synthetic biology certainly aspires to the creation of multi-cellular organisms. The re-creation of mammoths is envisaged. A "Pleistocene Park" has already been established in Siberia, where attempts are made to restore the mammoth ecosystem of the late Pleistocene period so that these mammoths will find a suitable habitat, alongside the woolly rhinoceros and others. What sort of ethics will be adequate here? And what about xeno-organisms, which are not part of the evolutionary tree of life at all? Anthropocentric ethics as we know it will hardly do justice to all of them.

---

[15]Or perhaps even of other planets: "One thing that was agreed upon, however, is the tremendous excitement and potential for synthetic biology to positively transform humanity in the coming years", as the Stanford-Brown iGEM team says, a group working at BioBricks parts which allow cells to survive harsh extraterrestrial conditions (Biobricks are defined DNA sequences with a known function). *Brown-Stanford iGEM 2011* (http://2011.igem.org/Team:Brown *Stanford/SynEthics/Summary*).

Jonas' fifth concern (1987, p. 48 f.) is that modern technology raises metaphysical issues, including the question as to the desirability of the continuation of the human species.[16] Synthetic biology is likely to accentuate further metaphysical issues. Traditional ontological categories seem suddenly insufficient. The distinction between nature and artefact becomes questionable if we can (artificially) create nature-like artefacts. (The recreated woolly mammoth would be an example of such an artificial nature-like organism.) Similarly, the category "life" becomes unclear. We have already seen how difficult it is to find any one common criterion that holds together all life-forms that we know– and perhaps their common evolutionary origin is the only one (cf. Toepfer, Sect. 4.5.). We are about to create "life" that is no longer part of the hitherto uninterrupted genealogy of some billion years. That is why Paul Thompson talks pithily about "the 'sexy' issues surrounding synthetic biology: issues that occasion widespread comment by philosophers, theologians and general pundits of science policy" (2012, p. 2).

Synthetic biology is thus obviously of high ethical relevance, and requires philosophical guidance. Let us therefore look more closely at the ethical issues it raises and ask in particular whether this new field requires a new ethics.

## 5.3  Ethical Issues of Synthetic Biology

### 5.3.1  Does Synthetic Biology Need a New Ethics?

Some authors argue that we are entering an area where regulations and principles once adequate in the field of genetic engineering no longer apply (Boldt and Müller 2008). Others maintain that a new ethics is not required (Eason 2012).[17] They argue that all relevant ethical issues have been already addressed in traditional bio-ethics. The Hastings Centre Report 2011, for example, "unanimously" concludes: "the field of synthetic biology does not require new regulation, oversight bodies, or a moratorium on advancing research at this time" (2011, p. 17). Some even warn against a "further balkanization of bioethics" by looking for new ethical principles (Parens and Johnston 2008, p. 1449).

It is yet early in the development of the science to come to a decision about the ethics. Synthetic biology has produced the first chemically synthesized yeast chromosome and the first bacterial genome, but the most promising potential benefits,

---

[16]When Craig Venter's group first made a synthetic bacterial genome, announced as the "First Self-Replicating Synthetic Bacterial Cell" (Press Release 20-May-2010), this achievement (and the way in which it was presented) stirred a debate on 'life'. Speculations on the metaphysical (and theological) implications of creating life followed (see, for example, the *Presidential Commission for the Study of Bioethics*, 2010).

[17]Perhaps not even for genetic engineering (see Block 2012).

and most worrisome risks, are not yet here.[18] Yet, because synthetic biology is such a dynamic enterprise, we can expect many more biotic artefacts in the near future (cf. Hümpel and Diekämper 2012). We can look at the products of synthetic biology and anticipate ethical challenges to come. Only then can we make efforts to consider and recommend safe development and control of this field for the greater good of all, before major problems arise. "The only reason for Ethics to lag behind this line of research is if we choose to allow it to do so" (Cho et al. 1999, p. 2090).

## 5.3.2 Risk

The problem of risk is well known from earlier debates in bioethics. It reaches, however, a new dimension with synthetic organisms with their specific charac-teristics. Most importantly, synthetic organisms are, because of their unparalleled depth of intervention on the gene pool, much less predictable. One might put this metaphorically: On the first two levels of synthetic biology, the words of the genetic code will be altered, on the level of Xenobiology, an entirely new genetic language is initiated. We know very little about the possible side-effects of these profound genetic changes, about the ecological traits of the new organisms (e.g. effects on soil food-web communities, other animals, plants, aquatic systems, and on sediments) or about their interactions with biological systems (habitat, target ecosystem, and other possible systems the new organisms can create). Neither is it clear to what extent they can transfer genes to other organisms or interact in patho-genic ways. In brief, because of the depth of intervention, the products of synthetic biology might behave in hitherto unknown and unpredictable ways.

The point of risk-assessment is to enable an evaluation of possible harm, or likelihood of harm occurring, after a release—but there is no available procedure or method for risk-analysis in synthetic biology. There are simply too many possi-ble interactions of synthetic organisms with their environment and too few we can anticipate. The familiarity principle no longer works, and, consequently, the tradi-tional methods of risk-analysis or technology-assessment are not able to deal with synthetic biology. We have little or no information on the probabilities of events occurring or on the magnitude of their effects. This marks a major dissimilarity with genetic engineering and explains why risk becomes such a pressing ethical problem.

That is true for all three levels of synthetic biology. We are ignorant of many fea-tures of novel life-forms on the level of engineering biology and know little about their chance of survival in different environments. Nor do we know much about

---

[18]There are other products occasionally discussed in literature (Church and Regis 2012): trans-genic "Glofish" with a fluorescent colour from a jelly fish, and goats which have a spider-gene so that spider silk can be extracted from their milk. These, however, are not proper examples of synthetic biology; they are not new, but only genetically modified life-forms.

the survivability, and potentially infectious nature, of synthetic proto-cells. At present, at least, it seems that the risks are highest in the world of proto-cell research and of engineering biology (Bedau et al. 2009), because deep interactions with the environment are likely here. After all, these organisms are designed to interact in some way with parts of the environment. As a consequence, new organisms and proto-cells might be toxic or infectious and may reproduce exponentially. On the level of xeno-life (organisms based on alternative biochemical structures) the danger of interaction with the environment is probably smaller. Xeno-life is genetically isolated and is unlikely to interfere with organisms with naturally-evolved DNA (Marliere 2009; Schmidt 2010), at least not directly. But it is still possible that some products of xeno-life might have harmful effects on the environment.

As long as synthetic organisms are confined within in the laboratory or in smaller settings, the risk might be limited, because the effects can be controlled and the organisms can be eliminated if necessary. But once these organisms are used for industrial purposes, such as fuel- or protein-production of some kind, huge quantities will be needed, and these are more and more difficult to control. Experience shows that some accidental release of organisms into the environment is inevitable. How might we assess the risk of a synthetic bacterium proliferating in nature? How might we calculate the probability of it replacing or contaminating other life-forms? Even if synthetic biology construct organisms with a reduced capacity to survive outside the laboratory or industrial process, such organisms might adapt rather quickly and interact in ways that are very different from what we have experienced so far.

Safety-related issues are wider than the accidental release of harmful synthetic biological agents or unintentional exposure to their products (pathogens, toxins, etc.). In particular *intended* misuse (this is the question of bio-security) is a significant risk in synthetic biology (Schmidt 2008; Bennett and Gilman 2009). A prime reason is that most of the relevant technologies are easily and cheaply available. It has already been shown that contagious viruses can be synthesised in laboratories by procuring DNA sequences online from commercial providers. The 1918 influenza virus could be isolated from exhumed corpses of people who have died of that disease; and corpses were frozen in permafrost soils in Russia, Norway, and Alaska (Tumpey et al. 2005). Entirely new viruses and other bio-weapons are, of course, possible. As a consequence, the 'dual-use dilemma' (important insights can be used in harmful or beneficial ways) poses a major problem. New technologies offer much potential for abuse. Bioterrorism (Bugl et al. 2007) is made the more possible by advances in engineering biology. At present proto-cells and xeno-life-forms seem much less useful as potential bio-weaponry.

To sum up, synthetic biology raises important ethical issues with regard to safety and security. These issues are not entirely new. In principle, the ethical challenge is the same as with genetic engineering: the crucial question is whether something we produce (an altered organism or a new life-form, process, or application) is safe enough to be developed and used, or whether it will present too great a risk for human beings, other animals, or the environment, whether directly or indirectly. There certainly is a continuity between the risk-problems of genetic engineering and synthetic biology; in both cases entirely new interactions are possible. But even if no new ethics is required, the classic methods of technology-assessment are not easily, or indeed no longer, applicable to many products of synthetic biology. That is why synthetic biology requires not only new methods of risk assessment but also extreme caution, at least until we have sufficient experience and until an adequate body of knowledge is gathered to allow proper evaluation and assessment. After all, the deliberate or accidental release of these organisms into the environment introduces potential ecological hazards of a hitherto unknown magnitude.

### 5.3.3 Global Justice

For a long time ethical debates within the synthetic biology community and among government policy-makers have dwelled almost exclusively upon risk (e.g. DFG et al. 2009), that is bio-safety and bio-security. This comes as no surprise: these are familiar challenges that we know from genetic engineering.

But the almost exclusive focus of the debate on risk, safety, and security has evoked reactions from civil society groups and philosophers (Boldt and Müller 2008; Torgersen 2009; Friends of the Earth 2012). They ask for a more profound ethical investigation of the field. Broader ethical issues, such as international justice, are raised.

Synthetic biology promises to create substitutes for natural products such as artemisinin, liquorice, palm oil, pyrethrin, rubber (isoprene), stevia, and vanillin (CSO 2011, p. 37f.). If these much-needed goods can be provided more cheaply and easily than the plant-based topical commodities that are currently still made by farming then this may deprive agricultural earners of their only income. Cheap production in itself is not unjust (the economic advantage of more efficient production is primarily a feature of the free market) but it raises issues of global responsibility when it has a negative impact on traditional exports and diminishes the livelihoods of farmers. All of these are serious issues. In 2011, the *International Civil Society Working Group on Synthetic Biology*[19] emphasised this concern and warned: "No inter-governmental body is addressing the potential disruptive impacts of synthetic biology on developing economies, particularly poor countries that depend on agricultural export commodities."

Ethical issues of global justice can also arise in connection with land use or pressure on biomass and its attendant socioeconomic consequences, especially in

---

[19]Consisting of the Action Group on Erosion, Technology and Concentration (ETC Group), Center for Food Safety Center for Food Safety Econexus, Friends of the Earth USA, International Center for Technology Assessment, and The Sustainability Council of New Zealand. Cf. http://www. econexus.info/sites/econexus/files/CSOsynbiosubmission_CBD_SBSTTA_synbio.pdf.

the developing world. Material that would otherwise be used for food or building activities or returned to the eco-system might become precious (as industrial-scale feedstock for biorefineries and fermentation tanks) and will be inaccessible to the poor. Even though all of these issues are not new, they are still relevant. These global concerns have already been widely discussed with regard to other new technologies such as genetic engineering and chemistry. (Raw materials from plant oils and waste animal fats supplied for biodiesel production raise similar problems.) Global concerns do indeed pose ethical questions, but they do not indicate a need for a new ethics of synthetic biology.

### 5.3.4 The 'Playing God' Argument

There is, however, a rather new type of ethical concern which often comes up in debates. Is synthetic biology 'playing God'? Though this question is not entirely unique to synthetic biology,[20] it is often mentioned in recent ethical debates on synthetic biology (e.g. Boldt et al. 2009; ETC. Group 2007; Balmer and Martin 2008; Bedau et al. 2009). The 'playing God' argument is also raised in discussions amongst lay people and in the mass media but notably hardly ever by theologians. It is "the curious fact", as Van den Belt rightly remarks (2009, p. 257) "that this argument is used mainly by secular organizations." Still, it is the most opaque argument in the debate. Let us therefore consider this objection and see what it might amount to.

A literal reading of the 'playing God' argument does not lead us very far; it is not clear within theological discourse why synthetic biology would amount to an illegitimate act of playing God. At least in the Christian tradition, human beings are generally seen as co-creators of the world, placed above all other creatures precisely because of this ability. Human beings were commissioned by God to improve the world, to build it up and even transform it. A distinction between two kinds of creative acts needs to be drawn. God is the only one who creates *ex nihilo* (*beri'ah*) in the first act of Genesis. There is also the creation out of pre-existing material (*yetzirah*)—that is God's work after the initial act and it is also the area for human creativity. This has been emphasized in the Christian as much as the Judaic tradition, which similarly sees human beings as co-creators and God's partners in the on-going process of creation. (It is also the origin of the understanding of human beings as creators in the Renaissance mentioned above). It was generally thought, therefore, appropriate that we should cultivate the earth, breed plants and

---

[20]As Peter Dabrock rightly observes (2009, 47): ""Playing God"—this reproach has accompanied modern biotechnology from its very beginnings. Almost every step forward in research has provoked vehement protest against the disregarding of creation: anaesthesia against pain, the birth control pill, transplantation medicine and diagnosing brain death, stem cell research and genetic engineering and many more innovations were faced with this reproach" For this history, Dabrock refers to Ramsey (1970), Chadwick (1989) and Coady (2009).

animals, create artworks, and build temples. Why should this *not* include the creation of biotic artefacts? Is this not just another realm of creative self-expression, of
human beings improving and transforming the world? Consequently, contemporary theologians have argued that there is nothing wrong in principle with creating
new life as long as it is not claimed that we do it better than God (or nature); synthetic biologists do not make such a claim (cf. Dabrock 2009, Haker 2012, Heavey
2013).

Let us therefore take the objection out of its theological context in order to
explore its other interpretations. Often it is merely mentioned as an expression of
angst or fear in face of synthetic biology, and sometimes the argument is straight-
forwardly dismissed as a pre-modern theological relic or conservative reservation
without argumentative value (see for example Dworkin 2000).This is probably the
reason why ethical analyses of synthetic biology tend not to address this argument.
But we should follow Anselm Müller (2004, p. 193) who points out that this 'argument' is, at least as often presented, ultimately a strong intuition. What could this
intuition be? It is, perhaps, the idea that we should not simply create all the things
we *can* create ('be like God') but rather respect certain *limits* when we create—
limits set for us as human beings with human capacities. Creating new life might
be a transgression of those limits: human hubris, as in the creation of the tower of
Babel. If this reconstruction is correct, then we could distinguish three aspects or
versions of the 'playing God' argument. Either we focus on the agents (I will call
this the Hubris-interpretation), or on the transgression of certain limits (the Taboo-
interpretation), or on the danger that comes with this transgression (the Sorcerer's
Apprentice-interpretation). The three versions probably exhaust the rationale
behind the argument in the bioethical debate. Let us look at them in turn.

- **The Hubris-interpretation**

The hubris-version sees a seductive danger in synthetic biology that is likely to
make the scientist overestimate his power. Driven further and further by what he
can do, simply because it is possible or "technically so sweet" (to quote
Oppenheimer again), he goes beyond his own capabilities or competence. Hubris,
here, refers to a state where the individual loses contact with reality in an immoral
way ("pride blinds") and acts unreasonably. This version of the argument can be
illustrated by Mary Shelley's *Frankenstein*, a novel which dwells upon the emotional horror of the young scientist and body-snatcher Dr Frankenstein when he
faces his own creation, a being assembled out of human parts from a graveyard.[21]
Out of similar hubris, it might be argued, synthetic biologists could also act foolishly, creating life-forms they cannot adequately deal with. (The recreation of
*Homo neanderthalensis*, advocated by George Church and others, might raise
problems that are similar to the ones described by Mary Shelley.[22])

---

[21]Mary Shelley portrays Victor Frankenstein as running away from his own creation, unable to
interact with it. Thus the tragic life of a lonely and rejected new creature begins.

[22]Church and Regis (2012), 137–140.

Synthetic biology seems to nurture this temptation particularly well. The danger of hubris can be seen as arising either from the actions caused by hubris or from the megalomaniac subject. If the focus is more on the action, then the critique concerns the dangerous outcomes of actions arising from hubris: "Pride goes before destruction, a haughty spirit before a fall" (Proverbs 16:18). In this version the hubris-argument comes close to the Sorcerer's Apprentice- interpretation which we will discuss below.[23] If the focus is the subject, then the concern is more with the deluded and inflated self-image and how this might lead to dangerous or immoral or amoral actions. Hubris can corrupt people's heart and cloud their judgment, as Raskolnikov in *Crime and Punishment* demonstrates. In a similar manner, synthetic biology might lead to the perversion of human beings. It might change attitudes towards nature by making us feel like sovereign masters of life. And it is quite possible that some of these attitudes are morally suspect.

Does, then, the 'playing God' objection in its hubris version show that a *new* ethics of synthetic biology is needed? Not necessarily. The peculiar seductive power of technology has been analyzed philosophically before the rise of synthetic biology. We find it already in Greek antiquity. Then Ovid re-tells in Book VIII of the *Metamorphoses* the story of Daedalus and his son Icarus. To escape from an island, Daedalus makes wings for himself and his son. Icarus takes these and flies higher and higher, as if to reach heaven. But the hot sun softens the wax, the feathers come off, and Icarus falls down and drowns. We learn from this myth that hubris (based on technological abilities) can make human beings act foolishly. And Heidegger (2002) points out that technology can affect our perception of the world quite generally by making us believe that everything, including nature, is at our (instrumental) disposal. Heidegger's point is certainly valid; technological artefacts do influence what we do (Illies and Meijers 2009) and even the way in which we perceive things and ourselves (cf. Peter-Paul Verbeek (2005) who coined the term "persuasive technology").[24] Long before synthetic biology people knew that technology can breed deluded and dangerous attitudes. Thus the hubris interpretation does not seem to demand a new ethics. It can, however, remind us of the danger of this new technology. The problem of human hubris should therefore certainly be part of the ethical debate around synthetic biology.

• **The Taboo-interpretation**

While the hubris-interpretation emphasises a perverted self-image, this second interpretation of the 'playing God' argument focuses on the act of creating life. Here the point is that by creating life synthetic biology does something that should not be done at all: it transgresses a limit that should be universally respected.

---

[23]A difference remains: the hubris-interpretation links the dangerous outcomes to a certain attitude of the agent while the Sorcerer's Apprentice-version is focusing primarily on the uncontrollable outcomes, independently from the dubious character of their creator.

[24]This influence has also been considered in the context of applied ethics, for example in healthcare. As Judith A. Erlen observes (1994, 66): "The promise of technology has the potential to act as a seductive force luring nurses and other health care providers to use it and luring patients to request it.".

This comes close to what Leon Kass has called the "'Yuck' factor".[25] In a famous article "The Wisdom of Repugnance" (1997), and in later writings, he argues that even within a secular society there are deeply rooted inhibitions against, for example, human cloning or incest. Any transgression is experienced as shocking and repulsive. Kass' central point is that we should take this strong emotional repulsion very seriously as an expression of a deep moral wisdom. This is a strong assumption: that there is reliable knowledge about moral limits discernible through our emotions and gut feelings.

But should we accept such boundaries, and if we choose to, why is synthetic biology seen as such a great transgression? Some ethical theories come to a similar conclusion, but by very different routes. Kantian ethics, most prominently, sees human value "beyond all price": a point of reference that is non-negotiable.[26] Very much in this spirit, some have argued that synthetic biology should not create humanoids or engineer the human genome (for example by trying to make human beings multi-virus resistant or change human cells so that they are cancer free[27]). Beneficial as the outcomes might be, the process would require using human beings for experiments and thus would instrumentalize them—something which, according to Kantian ethics, should be avoided at all costs.

In the debate on synthetic biology, the taboo-interpretation is often presented with a more general limit, namely "life" itself. After all, life has been a *factual* limit during most of history, in which people have exercised their scientific power and modified nature technically. Historically, life could never be made, it could only be destroyed; the step from life to death had been irreversible. That might be the reason why the factual limit was also seen as an ethical one; life seemed to be something holy and untouchable (Koepsell, p. 7). As Dabrock rightly observes (2009), synthetic biology "is on the point of breaking through a boundary deeply rooted in human cultural memory by penetrating into a domain that is believed to be exclusively reserved to the divine." As a consequence, 'life' and unique life-forms (such as species) have served as a yardstick for some ethical theories. In modern times, Albert Schweitzer (e.g. 2003) has argued that we must respect life in all its forms. Preston asserts that "environmental ethicists with a commitment to the normative significance of the historical evolutionary process may see synthetic biology as a moral 'line in the sand'" (Preston 2008, p. 23).

But why should the boundary of human creativity be here rather than there? One might argue that there are no good reasons to respect life as such (Illies 2006b) or,

---

[25]Cohen, Patricia (Jan 31, 2008). "Economists Dissect the 'Yuck' Factor". *The New York Times.* See also Fukayama and McGibben.

[26]Utilitarianism (an ethical theory of balancing pains and happiness or human preferences) has traditionally not offered any such boundary and was much criticised for the consequent problems. As a result, modern versions of Utilitarianism tend to include human freedom (and basic human rights) as such a boundary, thereby making the theory internally less consistent but its moral judgements more plausible.

[27]Cf. for example G. Church (http://www.spiegel.de/international/zeitgeist/george-church-explains-how-dna-will-be-construction-material-of-the-future-a-877634-2.html; 12.7.2013).

more pragmatically, that such a general moral demand to respect *all* life is of little practical use because it demands the impossible (Hauskeller 2006). (If, for example, *any* interference with life is regarded as an immoral act, then the breeding and the cultivation of plants and animals would be immoral.) Yet even if a consistent version of Albert Schweitzer's ethics, which allows for *some* interference, could be properly formulated, it is not clear why it should outlaw the creation of artificial organisms and the expansion of the diversity of life-forms. If every organism has a fundamental value then its destruction is wrong, as Schweitzer emphasises, then the creation of new life seems acceptable (or even laudable). (The risk of possible harmful side-effects of such new life remains, but that is germane to the Sorcerer's Apprentice-interpretation of the argument.) Thus bare 'life' as such is not a good candidate for the locus of absolute value, because such is not a basis for an ethical theory which works in practice.

Would a qualified notion of 'life' do a better job? Philippa Foot (2001) argues that forms of life provide their own measure of good and bad. A good dog, for example, is one that can walk and it is therefore bad to create one that cannot. Consequently, one might argue that creating a biotic artefact which could not have a normal life (for example, a dog without legs) would be objectionable, but the creation of a mammoth would not be.[28] But then, any such approach would have to show in detail what a normal form of life is; a particularly difficult task when it comes to artificial organisms. Is it, for example, improper for a goat to produce milk that contains spider-silk? What sort of arguments could be raised do determine what 'normal life' amounts to? All of this raises more questions than it answers.

Can, then the Taboo-interpretation of the 'playing God' argument show that we need a new ethics for synthetic biology? Any such conclusion faces two problems. It would have to show, firstly, that "natural life" can meaningfully be distinguished from "artificial life" (biotic artefacts). It must provide an argument why this difference is of high ethical relevance, so that any transgression, such as the creation of biotic artefacts, *should* be strictly forbidden. Otherwise the claim that we need a new ethics of synthetic biology *because* it transgresses a limit is begging the question: it presupposes that any such transgression is morally taboo and then 'concludes' that we must not allow it. But so long as we do not have strong arguments for the distinction between artificial and natural life (see Toepfer above, Sect. 4.5.), the Taboo-interpretation does not require a new ethics of synthetic biology.

• **The Sorcerer's Apprentice-interpretation**

This interpretation envisages the possibility of synthetic biology having incalculable and thus uncontrollable consequences, as Margret Engelhard has pointed out. The Sorcerer's Apprentice-interpretation is named after the poem by Goethe in which a magician's pupil uses his master's tricks but turns out to be unable to

---

[28]Philippa Foot's argument in *Natural Goodness* (2001) does not help. She claims that each species has its own standard as to how it should be treated, namely, in such a way that members of the species may realize their potential. This argument cannot say anything about the morality of the creation of new species.

control what he has created. The apprentice enchants a broom to fetch water in buckets—but he does not know how to stop it and so more and more water floods the floor. "Spirits that I've cited/My command ignore", he shouts in despair; but luckily his master appears and stops the broom. In this version, the 'playing God' argument emphasizes that we might not be able to control the outcome of the process. Does synthetic biology lead us to uncontrollable forces and thus to risks that are too great? Can synthetic biology guarantee that it will be able to control newly-created living organisms? After all, *qua* being alive, these new organisms potentially have a strong dynamic of growth, proliferation, and evolution. Like the waters flooding everywhere after the broom has been enchanted, we can imagine new organisms polluting our environment or displacing others organism, with no experienced sorcerer around to stop them.

In this interpretation, the argument touches on (and transgresses) the risk problems discussed above: synthetic biology might have uncontrollable effects. We have already discussed that it is nearly impossible to make, at present, a proper risk assessment; we simply know too little about the possible interactions of new life-forms with the environment. The sorcerer's apprentice reminds us of our methodological limitations in prediction. Like the apprentice, we are much better at bringing new things about than in controlling and limiting their effects.

There is, indeed, a serious problem when we act in ignorance of consequences; and this problem is substantial in the case of synthetic biology. We would still argue that this does not necessitate an entirely new ethics of synthetic biology. Jonas reminds us that our technological abilities are rather often not matched with an ability to anticipate (let alone control) global effects in the long run. (Extant examples include human-induced climate change, the problem of radioactive waste, and the damage caused by human-introduced non-native species into habitats.) One might, however, wonder whether this problem has reached a new magnitude in the case of synthetic biology. Let us therefore turn to "acting under uncertainty".

## 5.3.5 Acting Under Uncertainty

*Whenever* we act there is some uncertainty with respect to consequences. But acting 'under uncertainty' refers to the extreme case, where lies a serious ethical challenge. While 'risk' has to do with situations where probabilities can be assigned to possible harmful consequences, 'acting under uncertainty' refers to situations where no probability can be attributed to such consequences or where possible consequences are unknown. In these situations risks seem to be merely "calculable, hence controllable, islands in the sea of uncertainty" (Van Asselt and Vos 2006, p. 314). When we are in this sea, traditional risk-assessment is not of much help to decide what we should or should not do.

The economist Knight argued in 1921 that the two types of situation can be strictly distinguished, and that acting under uncertainty demands different strategies of decision-making from acting under risk. It does, however, seem that the difference between the two is not as strict as he imagines; and often the two are interwoven (van Asselt and Voss 2006, p. 313): we might know some, but not all, possible consequences of an action and might be able to make rough estimation as to their probabilities. The difference is also dynamic; over the time, our calculable knowledge of consequences can rise like a volcanic island out of the sea of uncertainty. This process is not linear (sometimes more knowledge will increase, rather than reduce, the amount of uncertainty by showing that our models were too simplistic) and may never have a terminus. Some uncertainties are 'radical' (Funtowicz and Ravetz 1990): either incalculable *in principle* (concerning, for example, human behavior in the future), or uncertain *in practice*, incalculable because of their complexity, given our current cognitive abilities and time-frames.

Let us specify what we mean by 'acting under uncertainty'. Hansson (1996, p. 369) distinguishes between four types of what he calls a "great uncertainty":[29]

He identifies firstly an "uncertainty of demarcation", when options are not really clear (1996, p. 370); either because we do not yet have a complete list of all possible options ("decision making with an *unfinished list of options*", p. 371), or we do not even know "what the decision is all about. It is not well determined what the scope of the decision is or what problem it is supposed to solve." (In the latter case he talks about "decision making with an *indeterminate decision horizon*" (p. 371)).

There is, secondly, an "uncertainty of consequences", when we do not know the consequences of different options. This can occur in two ways. Either we are ignorant of the probabilities of harmful consequences (all "we know about their probabilities is that they are nonzero" (p. 376)), or we do not even know what these consequences are (or, at least, do not know *all* consequences).

Thirdly, Hansson distinguishes "uncertainty of reliance" when the decision-maker is uncertain whether he can rely on the information of others (given by, for example, scientists).

There is, fourthly, a possible "uncertainty of values" when it is not obvious which values should be applied (Hansson 1996, p. 370).

What, then, are the relevant uncertainties in case of synthetic biology? Of the aspects listed by Hansson, the third does not require further discussion. There can, of course, be doubts as to whether we can trust scientific experts in synthetic biology; but there is no reason why this should be more troublesome than in the case of other technologies. Obviously, some people are less reliable than others, and self-interest influences judgment rather generally and not even consciously: "the violations of professionalism induced by conflicts of interest often occur automatically and without conscious awareness" (Moore and Loewenstein 2004, p. 199). This is a general truth about human nature and not specific to synthetic biology.

---

[29]Which he defines rath-er intuitively as "a situation in which the decision maker lacks much of the information that is taken for granted in the textbook cases" (1996, p. 369).

When we look at the fourth potential uncertainty, we find in practice little disagreement in the choice of most ethical norms. Ethical considerations of synthetic biology are generally based upon standard moral principles. Most people agree, for example, that the environment must be protected and that synthetic biology should not be applied to human beings. But the debate is still in its early days and it is easy to accept general values and prohibitions (such as non-application to human being) because synthetic biology is not able to act against these values. (As in the case of PID, disagreement may yet come, when technological capability increases.) Only in the context of the taboo-interpretation of the 'playing God' argument we can identify a straight-forward uncertainty of values: Is "life" a value that we should not touch at? There are also controversies about the hierarchy of values. Is a potential risk to the environment always a reason to limit the freedom of research? If not, where exactly should we draw the line? It is likely that there will be much discussion of this kind ahead; and it is desirable that there should be. But these issues are not new; we recognize them from the ethics debates concerning other technologies.

The uncertainty of demarcation and the uncertainty of consequences are obviously most relevant to synthetic biology. Let us look first at the problem of options. Synthetic biology is such a new field and the term covers so many different developments that we simply do not know what options there are. This ignorance is profound; the list of options is by no means complete and grows all the time. The number of DNA-sequences we can synthesize, for example, has grown enormously within a few years and the iGEM compiles an ever-growing list of genes which can be newly combined. In contrast to mere genetic engineering synthetic biology is not restricted to the evolutionary realm (see Chap. 2): the depth of intervention is much higher and so is the level of uncertainty. New technologies and options will emerge that we do not yet know of. In synthetic biology the scope of the decision has still to be clarified. Synthetic biology gives much rein to fantasy. Scientists claim that synthetic life creations could become tiny self-reproducing factories which perform many useful functions such as cleaning the environment, supplying us with limitless energy at low cost, or possibly even eliminating human sickness and age-related disease. There are few limits to the imagined benefits of this technology and thus few limits to our options. Imagination gives birth not only to great promise, but also to potential peril. We have critical cultural images and cautionary narratives about this new technology. Films such as the French-Canadian science fiction thriller *Splice* (2009) or Andrew Swann's *Moreau Series* (with its humanoid genetically-engineered animals) and other works of art, portray unwelcome options which might become available through synthetic biology. Both promise and peril remain to some extent still in the realms of fantasy, but such fantasy reveals how little we know about the options available to us.

The uncertainty of possible consequences is similarly high. This can be shown by reference to the 'checklist' Hansson suggests (1996, p. 378): a means of finding out whether we face a serious "high-level consequence uncertainty". Firstly, there must be an "asymmetry of uncertainty", by which he means that some new option has consequences much more uncertain than an alternative. Secondly, high-level uncertainty comes with the "qualitative novelty" of phenomena. Thirdly, where consequences have no obvious bounds in time or space uncertainty is increased. Finally we must face a possible "interference with complex systems in balance" (such as ecosystems) which we cannot bring back once disturbed. All these criteria are fulfilled by synthetic biology. The consequences of developing new synthetic life-forms are obviously much more uncertain than the option of *not* doing any further research in this area. The phenomena we are talking about are also qualitatively new and we do not know, therefore, much about them. Furthermore, the behavior of novel synthetic biological artefacts is inherently uncertain and unpredictable; but by being living biological entities they can also interfere with complex systems. Potentially, then, the products of synthetic biology could have an enormous or even devastating impact on biodiversity and thus on the functioning and resilience of ecosystems.[30]

This shows how urgently synthetic biology needs ethical analysis, even if we can apply the strategies that have been suggested to deal with uncertainty in general. There are two ethical rules which are crucial here. The first is to "diminish uncertainty by acquiring knowledge of the issue" (Tannert et al. 2007, p. 892). In the case of synthetic biology this amounts to the requirement of further research into the possible effects of biological artefacts. (This requirement is, however, extremely difficult to fulfill in practice. Possible effects are often too heterogenous and diffuse for systematic investigation.) The second is to apply a general precautionary principle in the absence of sufficient data, knowledge, or understanding. This principle requires that we do not increase danger without good reason, and thus do not find ourselves in situations of high uncertainty. In 2008 the *Conference of the Parties to the Convention on Biological Diversity* called at its ninth meeting for "submissions of information on synthetic biology and geo-engineering, while applying the precautionary approach to the field release of synthetic life, cell or genome into the environment" (Decision IX/29).[31] The precautionary principle deals with uncertainties in a proactive fashion (Jordan and O'Riordan 2000). If we cannot exclude that some option has a nonzero probability of extremely harmful consequence, and cannot perform any risk analysis, then we should not choose that option. This principle can come in weaker or stronger forms. A very strong form has been suggested by Ted Lockhart (2000): if we are not certain whether an action is morally permissible, and we do have an option we are certain is morally acceptable, then we should choose the latter (or, at least, that which most probably maximizes the good).

The precautionary principle should be applied when we can anticipate the possibility of great harm to the environment or to human beings (both those living and those yet to be born); and synthetic biology offers such possibility. A *caveat* might be that these harm scenarios must be reliable. We should take only "scientifically based harm scenarios" (Ekeli 2004, p. 431) into account, not *any* imaginable eventuality.

---

[30]Biodiversity loss is a pivotal driver of ecosystem change (cf. Hooper et al. 2012).

[31]Cf. http://www.cbd.int/cop9/doc/ (1.1.2014).

That sounds very reasonable, but the problem is, as expressed above, that we often do not know how to differentiate between these two kinds of scenarios.

### 5.3.6 New or Old Ethics?

There can be no doubt that synthetic biology forces us into an area of extreme uncertainty. We can justly describe this as a qualitative leap. The potential magnitude of possible consequences rightly disturbs us and requires a moral response. But, despite this seeming newness, most challenges of synthetic biology are familiar in principle, at least at present. They do not require a new ethics if 'new ethics' is understood as a field of new norms or values. The challenges are well known from previous contexts: genetic engineering is also facing major risks, acting under uncertainty, or issues of global justice.

However, synthetic biology is certainly radicalizing some of these challenges. As an extreme science—perhaps the apogee of modern science and technology—it might also be regarded as an extreme case of uncertainty. If we define 'new ethics' as a field that requires intense debate and new methods of Risk Assessment or Technology Assessment, then synthetic biology is in need of a new ethics, beyond the scope of traditional ethical debate.

## 5.4 The Need for New Ethical Debate and Its Problems

### 5.4.1 The Need for Debate

All the ethical issues of synthetic biology we considered thus far already exist in the arena of other modern technologies, in particular with genetic engineering. But even if synthetic biology does not raise new kinds of moral concern, it requires new answers to old concerns. After all, the uncertainty of what synthetic biology will bring about is much higher—and also the speed of its development. A lot of energy is put into this new science. Its enormous promise fascinates scientists and meets economic and political interests; in fields such as the production of new fuels or medicinal drugs, great hopes accompany the work of synthetic biology. These are areas of huge importance for industry but also for nation states, as the possibilities of, for example, bio-weapons, expand.

All of this explains why an ethical debate about where we want to go or not to go, a debate that will be more fundamental than those within genetic engineering, is much needed—"there is a critical need to examine the ethical implications of the new field of synthetic biology" (The Hasting's Centre 2008[32]). In other words,

---

[32]From Hasting's Center press information (September 11, 2008); cf. September 12 Issue of Science.

not a new ethics but a *new ethical debate* is required. Such a debate must be an enquiry into hitherto unknown fields in need of profound conceptual clarifications, a better understanding of the area and methods of dealing with uncertainty, and intense moral reflection. Only after such debate can we hope to find our way in a largely unexplored area with very specific challenges.

To be sure, this debate has already started, yet not many seem interested. Certainly not the public; only some ethicists, and merely few scientists are seriously engaged in this discussion. Why do so few people participate in the debate?

## 5.4.2 *The Lack of a Public Debate*

As it has been pointed out by Pardo and Hagen "most scientific developments take place silently, contributing to the continuous expansion of knowledge but known only to the corresponding subsets of the scientific community. They also tend to receive virtually no media attention"[33] What is true for most scientific developments is certainly the case for synthetic biology. Only very few people seem actively engaged in a public debate on it, and not many have even heard of it at all. 83 % of Europeans have no idea what synthetic biology is (Gaskell et al. 2010). Prado and Hagen add further reasons for this lack of a debate or even interest of the public.[34] One is that the potential applications of synthetic biology are not apparent to the public. We are still in the early days of this new technology. Synthetic biology and its products are not yet publicly visible, and neither its benefits nor its risks have reached the general awareness of the common person. In addition, synthetic biology is a rather complex field of technologies and scientific developments which can barely be understood without some scientific training and without a familiarity with basic genetic concepts. And they add that "cases of significant resistance to or open rejection of scientific developments by the public are an anomaly" (ibid.) and are thus not to be expected when the transformation is silent and more or less invisible. That is why many attempts to give the public a voice in these debates, and to encourage discussion and active engagement, including institutional mechanisms set up for this purpose, have failed.

## 5.4.3 *The Conceptual Challenge for Ethicists*

A further problem for this debate is that traditional ethical concepts (categories and norms) do not easily embrace the new phenomena. Synthetic biology has the potential to undermine many established moral responses. We have, for example, a

---

[33]See Sect. 6.4. in this book.
[34]Ibid.

basic environmental ethics which includes, amongst other things, the protection of biological diversity. But what does that amount to in the case of a new artificially created species? We referred to Hans Jonas above, who argues that modern technology shows us the limits of anthropocentric ethics—synthetic biology certainly does so and reveals the limits of many basic and generally-held assumptions and ethical concepts and categories.

Ethics is generally divided into three branches: Aplied Ethics, Normative Ethics, and Meta-Ethics. Applied Ethics is the area of actual problems; here general normative theories are applied to particular circumstances. The results can differ: there are conclusions that something is right or wrong, or conditional answers ('if you take standard y, than you have to do x').[35] Normative Ethics is the attempt to identify and justify an ethical theory which provides moral norms, rules, or guidelines. (Mill's Utilitarianims or Habermas' ethics of discourse are examples of Normative Ethics.) Meta-Ethics, the third branch, works on a more abstract level. It concerns itself with the analysis of basic concepts (What is the meaning of 'good'? Can it be reduced to usefulness?) and of methodological problems (How is an ethical norm justified?).

What is the impact of synthetic biology on these three branches of ethics? Some of the difficulties synthetic biology presents for *Applied Ethics* have been mentioned above. When dealing with new technologies, applied ethics has developed a set of standard procedures. Based upon the results of technology assessment, the risks and potential benefits of a new technology are examined, the public response and acceptance taken into account, and from this an ethical recommendation is formulated. Controversial areas remain, but the controversies are mainly between rival ethical theories and the different norms or values they apply. (Deontological ethics, for example, places a higher and even absolute value on human life, while utilitarian ethics allows for more 'trade-offs'.) Synthetic biology raises questions where standard procedures simply fail to provide answers. A prime example is risk-analysis in the face of much uncertainty. With respect to biotic artefacts we neither know what potential harm they might cause, nor can we specify the likelihood of such harm, and we can but speculate about potential benefit. In most cases, the complexity and novelty of synthetic biology does not allow for a traditional risk assessment. Another problem with Applied Ethics is the application of norms and values regarding 'life'. Can any ethical norm which demands respect for life be applied to biotic artefacts? Will the recreated woolly mammoth be an animal that commands respect?

Within *Normative Ethics* there arises the problem of new objects that are not covered by traditional normative theories. Most ethical theories have already evolved to the point where they can supply answers to moral problems concerning the environment, human interactions with (natural) species, and even with artefacts (such as artworks). But artificial organisms do not fit easily into any of the traditional theories. Is a synthetic organism a new species that should be preserved

---

[35]Some argue for an even more limited role of ethics, namely that it should confine itself to mediating between different positions and interests.

and protected from extinction? What of an organism with xeno-DNA? In his book *Animals, Humans, Machines. Blurring Boundaries* Mazis (2008) addresses this consequence of modern technology with regards to the difference between humans and animals. Synthetic biology goes even further and blurs more boundaries, like the one between natural and artificial (more so than, for example, nanotechnology or the genetic engineering of existing species).

One of the problems in Applied and Normative Ethics here is the need for a bridge between different and hitherto-unconnected ethical debates. Thompson (2012) has emphasized that the rather diffuse character of synthetic biology as a "platform technology" transcends traditional ethical theories. Such technologies, he argues, allow for "rapid and diffuse innovations and simultaneous product development in diffuse markets, often targeting sectors of the economy that have traditionally been thought to have little relationship to one another". As a consequence, he argues, we must bring different ethical debates together: "traditional medical applications", for example, and issues of global justice such as "land use and its attendant socioeconomic consequences, especially in the developing world" are all linked (Thompson 2012[36]).

At the Meta-Ethical level, these new technological developments stretch the boundaries of our traditional concepts. The notion of 'nature', for example, was often used in a normative way to indicate limits of intervention. Moral demands to preserve 'nature' are based upon the assumption that the concept places things with certain characteristics (living organisms, ecological systems) apart from inanimate things (minerals, artefacts). This intuition entails that everything belonging to 'nature' must be treated respectfully so that it can continue to exist in a (more or less) autonomous way. Synthetic biology sits close to the border between the living and the non-living, not only in its attempt to construct living organisms from non-living material, but also by constructing machines from living material. It abuts this border from both sides (see Deplazes and Huppenbauer 2009). Certainly, terms such as 'nature' or 'life' have already been under siege for some time, but synthetic biology seems to undermine them even more radically. The distinctions between natural and artificial, organism and machine, evolved and created, and between life and inanimate matter are now blurred, and even the concepts of identity and continuity face conceptual challenge.

There is, however, some disagreement about the importance of these conceptual changes: in particular about the implications of the changing concept of life. Are the ramifications of such conceptual changes crucial or irrelevant (e.g. Kaebnick 2009)? Will we simply get used to the idea that life can be synthetic and that human beings can create life, in the same way that we got used to seeing the world as a sphere (Tait 2009: p. 20)? Or does the mechanical view of life pose a threat to normative theories by implying that because life can always be recreated like a machine it is therefore ultimately worthless? In whatever way we answer these questions, they force us to reconsider some of our meta-ethical concepts.

---

[36]For this see also the 2010 report of the Presidential Commission for the Study of Bioethics.

## 5.4.4  The Silence of the Labs?

Scientists do discuss the ethics of synthetic biology, it seems. Any conference, book, or research application in the field has a section dedicated to ethical or social aspects of synthetic biology. It is, however, questionable whether all scientists are *seriously* involved in these debates. Some scientists, at least, regard them merely as compulsory exercises.[37] Ethical questions are not, of course, the most obvious concern for scientists.

There is, for one thing, a *motivational problem*. Most scientists acknowledge the need for reasonable risk-regulation (risks are, after all, quantifiable entities that can be dealt with in a roughly-speaking scientific manner), but otherwise are reluctant to engage in such debate. There are professional reasons for this lack of motivation. Following Ludwik Fleck, we can identify professional "thought-constraints" (1936, VI; 1935, IV.3) typical within the world of the natural scientist. In his work on the genesis and development of a scientific fact, Fleck has argued that groups of scientists form a specific "thought-collective" characterized by a collective "thought-style", which is a unique way of viewing and conceptualizing the world, but also includes shared practices. New members of the group are socialized into the specific thought-style by acquiring the worldview, concepts, and practices of the group. There are probably some norms belonging to the "thought-style" of the natural scientist, but they mostly concern how to do research (how to accumulate data, how to deal with the results of other scientists, etc.). Ethical issues, or debates on whole fields of research, are neither part of the academic education of scientists, nor do they play any role in the daily life of most laboratories. The style of working and thinking is shaped in a way independent of many ethical problems.

Even more profoundly, there are *scientific reasons* that seem to place ethical concerns outside the professional focus. In a Weberian sense, science is normatively neutral because rationality can only discover and analyze functional relations: it explains how and why things are as they are, but does not make claims about how they *should* be. Weber does not deny that we make normative claims (moral, aesthetic, and other), but for him they are based upon feelings, conventions, or religious convictions; they are not rationally established or justified. Science presents us with a disenchanted world, Weber argues, and ethics is part of the magical enchantment that we have overcome. Weber's understanding has become part of the fundamental scientific creed and explains the strong reservations many scientists have toward ethical debate beyond risk-analysis. Why should they rationally engage in debates which are, by their very nature in the scientist's view, not rational at all?

---

[37]One has the impression that some are hostile to any ethics, seeing ethics as a threat to science, as a set of obstructive restrictions from an ill-informed public or religious grouping. If such scientists can see any purpose in public debate, then it is as a means of creating a climate of confidence and to secure public funding for further research. Admittedly, this picture is a bit of a caricature. But the reality is sometimes not very sophisticated.

We should also note *institutional problems*. Work in a laboratory is extremely time-consuming and the competition between scientists is tough. There is no time to be wasted in ethical reflection or in the writing of papers which will not contribute to the scientist's professional development. Participation in ethical debates does not pay off and a scholarly article on the ethics of synthetic biology will normally not make its way into a natural science journal with a high impact factor. Most scientists can simply not afford to get engaged in such debate. But they should. They are the experts and know most about the technology and its potential effects. They are responsible for what they do.

## 5.4.5 Scientific and Public Language

In addition to the above-mentioned reasons and motives explaining why neither the public, nor ethicists, nor scientists become engaged in profound ethical investigation into synthetic biology, there are conceptual problems in communication (cf. e.g. Cho et al. 1999). Scientists and the public often use important terms in different ways. Consider, for example, the word 'nature'. Due to its non-normative approach, science uses 'nature' to mean a dynamic world of events governed by natural laws which can be expressed in mathematical language. Nature can be fully understood once we know all the mathematical formulae which describe the interdependencies between events. In contrast to the Aristotelian view, this understanding is strictly a-teleological: in nature, there are no underlying goals or ends. For the scientist our world is without purpose and governed by "blind" laws (Monod 1970). Investigating nature will therefore not unearth any ethical insights.

This interpretation of 'nature' differs enormously from that which dominated the West for nearly two millennia and still shapes public understanding today. "The poetry of the earth is never dead", as John Keats rightly states in his poem *On the Grasshopper and Cricket*. In the pre-Enlightenment view, nature is loaded with magic, with beauty, and even with normative content. Thus 'natural' implies good and is contrasted with 'artificial'. This understanding of the natural can affect what we do and what we choose not to do. Bernard Williams has coined the expression "thick concepts" for such "action-guiding" terms (1985, p. 140). The public understanding of 'nature' can ultimately be traced back to the classical view (of Aristotle and others) that nature is a universe of goal-directed entities, open to teleological (and, Christianity adds, theological) interpretations. Events in nature result from an inner dichotomy between change and being at rest; any explanation of movement will have to refer to this (Aristotle, *Physics* 2.1, 192b20–23). Outer forces affect nature only indirectly; the "active powers or potentialities (*dunameis*), which are external principles of change and being at rest (Aristotle, *Metaphysics* 9.8, 1049b5–10), [are] operative on the corresponding internal passive capacities or potentialities (*dunameis* again, *Metaphysics* 9.1, 1046a11–13)" (Bodnar 2012).

According to Aristotle, understanding nature is as much a metaphysical as a physical enterprise: it requires an inquiry into the true nature of things. Therefore we can always ask four question, Aristotle concludes, about any natural event, one of them being, 'For what purpose did it happen?' But this is a question explicitly rejected by science. Not only human beings and animals, but also plants and inanimate objects, are, in Aristotelianism, supposed to have an inner directedness, a *telos*. Understanding nature properly means understanding its point and purpose. It was this *telos* that could easily be linked to normative approaches, the full realization of one's *telos* being the good of one's existence. In this sense nature can be action-guiding, and it is still such an understanding which dominates public discourse on 'nature'. This discourse is much more Aristotelian than Galileian as Christina Aus der Au rightly remarks (2008).

Thus, the scientists and the public appear to speak of the same things but do not in reality do so. What is worse is that the two groups use the same words; scientists use widely-understood common language when they talk about 'nature', but, in the eyes of the public, without taking the normative connotations seriously. This is at least one reason for the lay person's suspicion of the scientist. And for most scientists, lay people are relics of a pre-scientific worldview which needs to be eradicated. This, in turn, explains, why many lay people do not feel that they are being taken seriously and therefore distrust scientists (Aus der Au 2008, p. 24). What has been said of the term 'nature' is also true of other key concepts such as 'life' and 'artificial'. These are used as a "thick concepts" by many; as notions with normative implications. But not by scientists.

The problems, then, encountered here are not only the conceptual difficulties presented by new technology, but also professional, conceptual, and institutional inhibitions on the part of the participants. They must be overcome for substantial answers to be reached. What is necessary is a real encounter between scientist and public, held in a mutually intelligible *lingua franca*, and an overcoming of professional inhibition. This requires ethical reflection, political effort, economic and scientific incentives, as much as a willingness on both sides to enter the debate seriously. The lay person must try to understand what is going on in synthetic biology, and the scientist must show the same cooperativeness. The scientist must try to empathise with those who express fear or misgiving, and also understand that, and how, they use the same terminology differently from them. We hope that some of the clarification achieved here will be of help in this cause.

But we need more. We need actual suggestions as to where the debate can start from. The point of an ethical debate on synthetic biology is to formulate answers: to find out whether society has good reasons to promote, allow, or disallow research areas, products, developments, or individual technologies within the world of synthetic biology. In good Socratic tradition (of setting up a hypothesis which can be argued against) I will end this chapter by making a moral proposal for the level of engineering biology. I suggest what might be called 'The Three Moral Laws' of synthetic biology.

## 5.5  The Three Moral Laws

### 5.5.1  A Moral Proposal

In his 1942 short story *Runaround*, Isaac Asimov, the author of several science fiction stories, introduced a set of three laws or rules which any robot must obey. These "Three Laws of Robotics" (often simply called "The Laws") read as follows:

1. "A robot may not injure a human being or, through inaction, allow a human being to come to harm."
2. "A robot must obey the orders given to it by human beings, except where such orders would conflict with the First Law."
3. "A robot must protect its own existence as long as such protection does not conflict with the First or Second Law."

Following a suggestion by Michael Bölker, we can try to apply the Three Laws to synthetic biology, in particular to the level of engineering biology where we will soon find artificial organisms which can interact easily with the environment, at least in principle. These "Three Laws of Synthetic Biology" read:

1. A biotic artefact may not injure a human being.
   Synthetic biology must obey a precautionary principle and must avoid causing any significant immediate or long-term risk to human beings or the environment.
2. A biotic artefact must be strictly functional, except where this would conflict with the First Law.
   Biotic artefacts must be designed for a limited purpose. Synthetic biology must not create artefacts for their own sake nor artefacts with a possibility for active further development (something we might call evolutionary autonomy).
3. A biotic artefact must be protected and we should respect it as a form of life as long as such protection does not conflict with the First or Second Law.
   The life of biotic artefacts must be respected and their value acknowledged but only to the extent that they do not threaten mankind, and where they are strictly functional. We must not respect or support their evolutionary autonomy.

Obviously, it was easier for Asimov than for us to formulate his Three Laws and to hope for their general acceptance. As a writer he is the creator of his own (literary) universe, where he sets the rules and determines what happens: his rules are for robots and can directly be incorporated into them. They cannot be bypassed, even if the robots might act in strange and unexpected ways when they apply the laws (that is the punch line of many of Asimov's stories). Things in the real world, especially products of synthetic biology, differ from entities in Asimov's universe. For a start, moral laws cannot be programmed into biotic artefacts; we must address moral issues to the synthetic biologists, lawgivers, and other stakeholders. Furthermore, we cannot determine how such people will behave, because we are not the authors of their lives and actions. We relate to them as free, independent beings. That is why we can merely present the laws and argue for them, hoping that people will be motivated to act upon them. Let us, therefore, say a little more about the Three Laws of Synthetic Biology and why we *should* accept them.

## 5.5.2  First Law: A Biotic Artefact May not Injure a Human Being

Most ethical theories accept human life (and its necessary conditions) as valuable in themselves. The respect for human beings, their autonomy and well-being are generally acknowledged norms. And theories in the Kantian tradition see human dignity as the fundamental value in all ethics. While traditional ethics focuses on human beings who already exist, Jonas expands the principle to include those who have not yet been born. His moral imperative of responsibility (1984, p. 43) reads: "Act so that the effects of your action are compatible with the permanence of genuine human life."

Kantian ethics sees human value "beyond all price", as an absolute which must not be subject to any weighting or trade off. A contemporary version of this can be found in the work of Roger Brownsword, who argues that there is an absolute which all ethical theories must acknowledge; namely the conditions necessary for any moral community to exist. This follows from the requirement that moral demands be self-consistent and do not undermine themselves. If as a society we look for moral orientation, we need to identify all the conditions necessary for this society to exist as a moral community. These conditions must be respected at all costs, in order to avoid theoretical inconsistency and practical self-destruction. The moral order must be geared towards longevity and may not include norms whose consequences turn against that order. Human dignity is a good candidate as an absolute. If we fail to treat human beings with respect, we will certainly undermine the possibility of any moral community.

If we take a fundamental norm like 'secure the conditions necessary for any moral community to exist' as a starting point of ethical reasoning, the First Law follows. When creating biotic artefacts, synthetic biology must always avoid harming human beings. This requires not merely the avoidance of obvious harm (as might be caused by the creation of, for example, a dangerous virus) but also the prohibition of any form of experimentation on humanoids. Re-creating Neanderthal Man (even if it is "technically sweet") seems clearly against this law. Not only will such an attempt be unavoidably accompanied by the misery of many failed attempts, the artificial *Homo neanderthalensis* will, on a purely biological basis, not be well equipped to withstand contemporary diseases, bacteria, or viruses; so his life might be short and full of suffering. Moreover, we know nearly nothing about his needs, wishes, or desires, and what might happen when he finds himself in a situation where he is a lonely alien, in a world he was not made for, would be profoundly unpredictable. Jürgen Habermas' argument against cloning (namely that an asymmetrical situation is created where the clone has limited freedom) applies here in an even more extreme way. It is hard to see how a Neanderthal Man could become a truly autonomous being and a part of our moral community. The problem is not the creation of a potentially autonomous being *as such* (after all, we do it by sexual reproduction all the time), but the circumstances that prevent the being actually developing this autonomy. Autonomy, after all, is the result of growing up amongst kin in a supporting culture; things he will not experience.

The First Law also dictates that we must be extremely careful with respect to possible risks. If human existence is the fundamental value, then nothing could justify risking human existence as such in the short or long term. If any biotic artefact presents a possible risk to human beings, directly or indirectly, or by affecting the environment in negative ways, the burden of proof that it is *not* harmful falls on synthetic biology. Research in synthetic biology must always be accompanied by careful investigation of possible consequences, with sober multi-disciplinary risk-assessment. Only such assessment can be a fair basis for ethical debate on action and policy. At present, however, we seem unable to execute any proper risk-assessment for most products of synthetic biology. And as argued above, acting under uncertainty requires a strict precautionary principle. Obviously, it is difficult to determine how strict the precautionary approach should be: does it mean that creating *any* new artefact must be outlawed, simply because we cannot know all consequences in advance? The First Law certainly demands extreme care at least when there is potential harm to human beings.

### 5.5.3 Second Law: A Biotic Artefact Must Be Strictly Functional

The Second Law can be deduced from the First Law. To demonstrate this, let us review what we have argued concerning the First Law: we claimed that the First Law implies that there should be no significant risk of harm to human beings. This statement, however, requires further qualification. Not to run *any* risk would be equivalent to prohibiting synthetic biology completely because a minimal level of risk will always obtain. After all, we live in a world of probabilities, where we can never be absolutely certain about the outcomes of events. (And as Chaos Theory tells us, in a non-linear system, as most physical systems are, simple changes can produce complex effects.) Hence, if we followed the strict principle of avoiding any risk whatsoever, no new technology would be acceptable. The laptop on the table might electrocute its user; and if we take up our long-forgotten fountain pen instead, it is still possible that something dreadful happens. But strict risk-avoidance is also morally problematic: it might imply absolute passivity, which is not an ideal of the moral life. Moreover, even passivity might be risky. We should accept instead that we live in an imperfect world where everything we do has some risk of bad (yet unintended) consequence. Absolute risk-avoidance is not a possible principle; there are good reasons to accept a principle of minimal risk instead.

It follows that we must make some benefit/risk assessment; possible harm and its probability must be assessed against the benefit of an action or new technology. But as stated above, any such calculation will be extremely complex and difficult in case of synthetic biology. We simply know too little about the likelihood of genetic pollution, direct or indirect effects on other organisms or the environment, etc. But we do know something about the relative importance of certain features. Let us look at them:

Firstly, we know that there must be a clear non-negligible benefit in the technology; otherwise no risk would seem acceptable. If there are no potential benefits any risk would be too high, simply because the minimal potential harm is not counterbalanced by any benefit. We can also conclude that, if there are two biotic artefacts A and B, which are alike with the exception that A has a beneficial feature while B has a merely neutral ones, then we know that it is ethically better, *ceteris paribus*, to create A rather than B. And as B has no beneficial feature, we know a priori that the risk will outweigh the benefits and that the risk/benefit balance will weigh against its creation. A biotic artefact *must* promise some particular benefit and be designed for a clear purpose; otherwise we should not countenance its creation. This purpose should be strictly limited. The artefact must not have an unpredictable variety of features and abilities which are neither needed for its existence, nor are potentially beneficial for humans. Any additional feature, over and above the clear purpose of creation, will increase the risk of unforeseeable negative interactions with the environment. The ideal (and actually the only acceptable) biotic artefact is a minimal organism with a clearly defined and limited function. (This criterion is likely to be in harmony with economic interests. Newly created organisms are designed to fulfill a specific task within industrial scale operations, and designed not to lose or alter their ability to perform this task. In most cases, evolutionary changes in the organism are not welcomed.)

Biotic artefacts should not have the potential for further evolutionary development. Certainly, by being a living organism such an artefact will always be, in principle, capable of variation and adaptation and thus a possible subject for natural selection. Some evolutionary potential cannot be excluded. But this potential must be minimized, if possible, because it increases the likelihood of uncontrollable and unwanted harm. If the artefact becomes evolutionary active, it will replace other forms of life (as implied by the term "natural selection")[38] and thereby increase its role in the ecosystem, which makes it yet more difficult to control. Again, this Second Law can be argued for in an a priori fashion without attaching

---

[38]Replacement happens on two levels: a better-adapted strain of one species replaces other strains of that same species, and successful species can replace other species. In the case of biotic artefacts, the first is less dangerous than the second, but even the first is potentially harmful: it means that the new life form may become better and better adapted to its environment and therefore more and amore able to interact with it.

any absolute value to risks or benefits. If there are two biotic artefacts A and B, which are alike with the exception that B has more evolutionary potential than A, then B has the potential, in the long run, to interact more intensively with the environment, replace more life-forms, and change things more effectively than A. Since all these influences are potentially harmful, B itself is potentially more harmful than A. It follows that it is ethically better, *ceteris paribus*, to create A than B. Another consequence is that synthetic biology should not explore areas where functional applications are not immediately apparent. Research without a clear goal ('curiosity-driven science', as it is sometimes called) is not acceptable as long as we follow the pre-cautionary principle (even if many unanticipated scientific breakthroughs are only possible if this freedom of research were granted). At present, it seems the most rational (and thus ethical) principle that biotic artefacts must be designed for a strictly limited purpose and must not have the potential of dynamic (evolutionary) development.

### 5.5.4  Third Law: A Biotic Artefact Must Be Protected

This law is based upon the idea that all forms of life, whether biotic artefacts or traditional life-forms, deserve respect. Observance of this law is qualified by the First and Second Laws: only un-harmful and strictly functional artefacts are supposed to be created and only they deserve the minimal respect commanded by the Third Law. If a biotic artefact has too much evolutionary potential, it violates the Second Law and therefore must not be created, being simply too dangerous for a world such as ours.

Why should this Third Law be applied? We have argued above against the idea that 'life' as such is an absolute value, mainly because it is not the possible basis for an ethical theory. The Third Law is different because it is based on a principle of fairness. We do in practice give some normative status to plants or animals: we regard it as immoral to make a species extinct and consider cruel treatment of at least higher animals to be immoral. But then, we should pay the same respect to members of artificial species. Fairness would demand that we treat a higher artificial animal, such as a mammoth, in the same way that we would treat, say, an elephant. They deserve the same respect we owe to all higher animals: we must not make them suffer, and we must keep them in appropriate conditions—as long as the First and Second Laws are obeyed.

In most cases such respect for a novel form of life, based upon this fairness principal, does not have much 'cash value': it does not entail that the organism in question must be protected under all circumstances (cf. Boldt et al. 2009). This is because we are, in practice, mainly dealing with very simple organisms (such as proto-cells and bacteria) towards which we seem to have few, if any, obligations. Even 'natural' bacteria do not seem to ask for much respect, so the fairness-principle does not create many obligations in the case of artificial bacteria.

## 5.5.5 Short Conclusion

What follows? The three Laws are a suggestion which may at least generate debate; only in the outcome of such debate might they be shown to be plausible compasses for moral orientation in the field of synthetic biology. It would be good to have such a debate before the developments of synthetic biology urgently demand ethical or political answers, because by then it might be too late to give any.

# References

Aus der Au C (2008) Der Begriff der Natur in der Debatte über die grüne Gentechnik. In: Busch RJ, Pütz G (eds) Biotechnologie in gesellschaftlicher Deutung, Utz, München, pp 21–28

Bacon F (1711) Meditationes Sacrae 1597. In: The works, with several additional pieces never before printed in any edition of his works. to which is prefixed a new life of the author by Mr. Mallet, vol II, London, pp 402–403

Balmer A, Martin P (2008) Synthetic biology. Social and ethical challenges. An independent review commissioned by the Biotechnology and Biological Sciences Research Council (BBSRC). http://www.bbsrc.ac.uk/organisation/policies/reviews/scientific-areas/0806-synthetic-biology.aspx. Accessed 2 July 2013

Bedau MA et al (2009) Social and ethical checkpoints for bottom-up synthetic biology, or protocells. Syst Synth Biol 3:65–75

Bennett G, Gilman N (2009) From synthetic biology to biohacking: are we prepared? Nat Biotechnol 27(12):1109–1111

Block WE (2012) Synthetic biology does not need a synthetic bioethics: give me that old time (Libertarian) ethics. Ethics Policy Environ 15(1) (available at SSRN: http://ssrn.com/abstract=2032738)

Bodnar I (2012) Aristotle's natural philosophy the Stanford Encyclopedia of Philosophy (Spring 2012 Edition) (http://plato.stanford.edu/archives/spr2012/entries/aristotle-natphil/)

Boldt J, Müller O (2008) Newtons of the leaves of grass. Nat Biotechnol 26(4):387–389

Boldt J, Müller O, Maio G (2009) Synthetische Biologie. Eine ethisch-philosophische Analyse. BBL, Bern

Brown D (2012) Navigating the perfect moral storm: climate ethics. Earthscan, London

Bugl H et al (2007) DNA synthesis and biological security. Nat Biotechnol 25(6):627–629

Burckhardt K (1878) Die Cultur der Renaissance in Italien, Basel

Caplan A (2010) The end of vitalism. Nature 465:423

Chadwick R (1989) Playing God. Cogito 3:186–193

Cho MK, Magnus D et al (1999) Genetics: ethical considerations in synthesizing a minimal genome. Science 286(5447):2087–2090

Church G, Regis E (2012) Regenesis. How synthetic biology will reinvent nature and ourselves. Basic Books, New York

Coady CAJ (2009) Playing God. In: Savulescu J, Bostrom N (eds) Human enhancement. Oxford University Press, Oxford, pp 155–180

Cockell CS (2005) The value of microorganisms. Environ Ethics 27(4):375–390

Connor MR, Atsumi S (2010) Synthetic biology guides biofuel production (review article). J Biomed Biotechnol 2010. CSO SynBio submission to CBD SBSTTA Oct 2011. (http://www.econexus.info/publication/potential-impacts-synthetic-biology-biodiversity). Accessed 13 Dec 2013

Dabrock P (2009) Playing God? Synthetic biology as a theological and ethical challenge. Syst Synth Biol 3(1–4):47–54

Darwin C (1993) The autobiography of Charles Darwin (Barlow N ed). W.W. Norton Co, United States

DeMarco D (1986) The dispute between Galileo and the Catholic Church. Homilet Pastor Rev CI 3(23–51):53–59

Deplazes A, Huppenbauer M (2009) Synthetic organisms and living machines: positioning the products of synthetic biology at the borderline between living and non-living matter. Syst Synth Biol 3(1–4):55–63

Dupuy J-P (2009) Technology and Metaphysics. In: Olsen JKB, Pedersen SA, Hendricks VF (eds) A companion to the philosophy of technology. Blackwell, New York, pp 214–217

Dworkin R (2000) Sovereign virtue. The theory and practice of equality. Harvard University Press, Cambridge

Eason RE (2012) Synthetic biology already has a model to follow. Ethics Policy Environ 15(1):21–24

Ekeli KS (2004) Environmental risks, uncertainty and intergenerational ethics. Environ Values 13:421–448

Erlen J (1994) Technology's seductive power. Orthop Nurs 13(6):5–58

ETC Group (2007) Extreme genetic engineering. An introduction to synthetic biology. http://www.etcgroup.org/content/extreme-genetic-engineering-introduction-synthetic-biology. Accessed 27 April 2013

Foot P (2001) Natural goodness. Clarendon Press, Oxford

Friends of the Earth (2012) Annual report. http://www.foei.org/wp-content/uploads/2013/12/Annual-report-2012.pdf

Funtowicz SO, Ravetz JR (1990) The emergence of post-normal science. In: von Schomberg R (ed) Science, politics, and morality: scientific uncertainty and decision making. Kluwer, Dordrecht, pp 85–123

Galilei G (1842) Opere Complete di Galileo Galilei, Florence

Gaskell G et al (2010) Europeans and biotechnology in 2010. Winds of change? European Commission

Grunwald A (2009) Technology assessment: concepts and methods. In: Dov M, Meijers A, Woods J (eds) Handbook of the philosophy of science. Philosophy of technology and engineering sciences, vol 9. Elsevier, Amsterdam, pp 1103–1146

Haker H (2012) Eine Verhältnisbestimmung von Theologie und Synthetischer Biologie aus ethischer Sicht. In: Köchy K, Hümpel A (eds) Synthetische Biologie. Entwicklung einer neuen Ingenieursbiologie? Themenband der interdisziplinären Arbeitsgruppe Gentechnologiebericht. Forum W, Dornburg, pp 195–213

Hansson SO (1996) Decision making under great uncertainty. Philos Soc Sci 26:369–386

Hansson SO (2009) Technology, prosperity, and risk. In: Olsen JKB, Pedersen SA, Hendricks VF (eds) A companion to the philosophy of technology. Blackwell, New York, pp 484–494

Hauskeller M (2006) Verantwortung für das Leben? Schweitzers Dilemma. In: Hauskeller M (ed) Ethik des Lebens. Graue Edition, Zug, pp 210–236

Heaf D, Wirz J (eds) (2002) Genetic engineering and the intrinsic value and integrity of plants and animals, work-shop, royal botanic garden. Edinburgh. Ifgene, Dornach, pp 5–10

Heavey P (2013) The place of God in synthetic biology: how will the Catholic Church respond? Bioethics 27(1):36–47

Heidegger M (2002) Die Technik und die Kehre. Cotta, Stuttgart

Hösle V (1991) Philosophie der ökologischen Krise: Moskauer Vorträge. Beck, München

Hooper D et al (2012) A global synthesis reveals biodiversity loss as a major driver of ecosystem change. Nature 486:105–108

Hümpel A,Diekämper J (2012) Daten zu ausgewählten Indikatoren. In: Köchy K, Hümpel A (eds) Synthetische Biologie. Entwicklung einer neuen Ingenieursbiologie? Themenband der interdisziplinären Arbeitsgruppe Gentechnologiebericht. Forum W, Dornburg, pp 264–265

Illies C (2006a) Philosophische Anthropologie im biologischen Zeitalter. Suhrkamp, Frankfurt a.M

Illies C (2006b) Ehrfurcht statt Begründung? Albert Schweitzers Versuch einer Grundlegung der
    Ethik. In: Hauskeller M (ed) Ethik des Lebens, Graue Edition, Zug, pp. 189–209
Illies C, Meijers A (2009) Artefacts without agency. Monist 92(3):422–443
Jonas H (1984) The imperative of responsibility. UCP, Chicago
Jonas H (1987) Technik, Medizin und Ethik. Suhrkamp, Frankfurt a.M
Jordan A, O'Riordan T (2004) The precautionary principle: a legal and policy history. In:
    Martuzzi M, Tickner JA (eds) The precautionary principle: protecting public health, the envi-
    ronment and the future of our children. World Health Organization (1), pp 31–47
Kaebnick GE (2009) Synthetic biology: engineering life. Lahey Clin J Med Ethics 16(3):6–7
Kant I (1790) Kritik der Urteilskraft. Akademie Ausgabe Vol. 5. De Gruyter, Berlin 1968
Kass L (1997) The wisdom of repugnance. The New Republic (2 June 1997), pp 17–26.
Knight FH (1921) Risk, uncertainty and profit. Houghton Mifflin, Boston
Lee K (2009) Biology and Technology. In: Olsen JKB, Pedersen SA, Hendricks VF (eds) A com-
    panion to the philosophy of technology. Blackwell, New York, pp 99–103
Lockhart T (2000) Moral uncertainty and its consequences. OUP, Oxford
Löwith K (1986) Vicos Grundsatz: verum et factum convertuntur. Seine theologische Prämisse
    und deren säkulare Konsequenzen. In: Sämtliche Schriften, Band 9. Metzler, Stuttgart,
    pp 195–227
Marliere P (2009) The farther, the safer: a manifesto for securely navigating synthetic species
    away from the old living world. Syst Synth Biol 3(1–4):77–84
Mazis GA (2008) Animals, humans, machines. Blurring boundaries. Suny Press, New York
Mitcham C, Waelbers K (2009) Technology and ethics: overview. In: Olsen JKB, Pedersen SA,
    Hendricks VF (eds) A companion to the philosophy of technology. Blackwell, New York, pp
    367–383
Monoid J (1970) Le hasard et la nécessité. Essai sur la philosophie naturelle de la biologie mod-
    ern. Le Seuil, Paris
Moore D, Loewenstein G (2004) Self-interest, automaticity, and the psychology of conflict of
    interest. Soc Justice Res 17(2):189–201
Müller A (2004) Lasst uns Menschen machen!. Kohlhammer, Stuttgart
Na J (2002) Praktische vernunft und geschichte bei vico und hegel. Königshausen und Neumann,
    Würzburg
Parens E, Johnston J et al (2008) Ethics: do we need "synthetic bioethics"? Science
    321(5895):1449
Preston C (2008) Synthetic biology: drawing a line in Darwin's sand. Environ Values
    17(1):23–39
Ramsey P (1970) Fabricated man. The ethics of genetic control. Yale University Press, New
    Haven
Rawls J (1971) Theory of justice. Harvard University Press, Cambridge
Schmidt M (2008) Diffusion of synthetic biology: a challenge to biosafety. Syst Synth Biol
    2(1–2):1–6
Schmidt M (2010) Xenobiology: a new form of life as the ultimate biosafety tool. BioEssays
    32(4):322–331
Schweitzer A (2003) Die Ehrfurcht vor dem Leben. Beck, München
Simmel G (1900) Philosophie des Geldes. Suhrkamp: Frankfurt a. M. 1994
Tannert C, Elvers H-D, Jamdrig B (2007) The ethics of uncertainty. EMBO Rep 8(10):892–896
Tait J (2009) Upstream engagement and the governance of science. The shadow of the geneti-
    cally modified crops experience in Europe. EMBO reports 10, S18–S22
Thompson P (2012) Synthetic biology needs a synthetic bioethics. Ethics Policy Environ
    15(1):1–20
Torgersen H (2009) Synthetic biology in society: learning from past experience? Syst Synth Biol
    3(1–4):9–17
Tumpey et al (2005) Characterization of the reconstructed 1918 Spanish influenza Pandemic
    Virus. Science 310 (5745):77–80

Van Asselt M, Vos E (2006) The precautionary principle and the uncertainty paradox. J Risk Res 9(4):313–336

Van Den Belt H (2009) Playing God in Frankenstein's footsteps: synthetic biology and the meaning of life. Nanoethics 3(3):257–268

Verbeek P-P (2005) What things do. Philosophical reflections on technology, agency, and design. Pennsylvania State University Press, Philadelphia

Vico G (1710) De antiquissima Italorum sapientia, ex linguae latinae originibus eruenda. Nabu Press 2011

Weber M (1934) Die protestantische Ethik und der Geist des Kapitalismus. Mohr, Tübingen

White L Jr (1978) Medieval religion and technology: collected essays. University of California Press, Berkeley

White EB (2007) Letters of E.B. White. Harper Perennial; Reprint edition

Williams B (1985) Ethics and the limits of philosophy. Collins, London

Winner L (1980) Do artefacts have politics? Daedalus 109(1):121–136

# Chapter 6
# Synthetic Biology: Public Perceptions of an Emergent Field

Rafael Pardo Avellaneda and Kristin Hagen

**Abstract** We analyze some of the issues that synthetic biology raises for the social sciences within the "public perceptions of science" framework. The changing roles of public perceptions in policy making are described in relation with changes in the institutional and cultural contexts of science. We take a closer look at the available empirical evidence about public views on synthetic biology against the background of what is known about public perceptions of biotechnology more generally. Many vectors influence public attitudes to biotechnology, notably risk perceptions, tradeoffs between goals and means, ethical views, and trust in science and regulatory institutions. Attitudes are also associated with frames, symbols and worldviews. One of the central worldviews that affects subsets of the life sciences is the current vision of nature: many people are aware of problematic aspects of economic growth that makes intensive use of science and technology, and there is therefore sensitivity to scientific progress that further challenges the boundaries of "natural" processes and objects. Synthetic biology has components in potential conflict with the public's preference for "naturalness" in many areas, although this is at present dormant due to the low salience of synthetic biology in the media and public.

R. Pardo Avellaneda (✉)
Fundación BBVA, Paseo de Recoletos 10, 28001 Madrid, Spain
e-mail: rpardo@fbbva.es

K. Hagen
EA European Academy of Technology and Innovation Assessment GmbH,
Wilhelmstr 56, 53474 Bad Neuenahr-Ahrweiler, Germany
e-mail: kristin.hagen@ea-aw.de

© Springer International Publishing Switzerland 2016                     127
M. Engelhard (ed.), *Synthetic Biology Analysed*, Ethics of Science
and Technology Assessment 44, DOI 10.1007/978-3-319-25145-5_6

## 6.1  Introduction

Synthetic biology is an emergent scientific area with an explicit dual purpose: to provide a new and deeper understanding of biological systems, and to design from scratch biological parts and systems for a number of practical purposes, from medicine to the restoration of the natural environment. In contrast to most scientific areas, in which a significant accumulation of knowledge takes place before applications are envisioned, leading figures in synthetic biology have publicly portrayed a future of multiple innovations and applications, despite the fact that few scientific breakthroughs have so far been produced. Indeed the amount of hype and articulation of grandiose visions has been considerable by any standards. In this respect, synthetic biology reflects the new rules of the game for doing science: to claim critically important practical benefits in order to achieve rapid recognition, advancement in academia and access to custom-made support programs from governments and private investors. A more significant novelty of this still fuzzy field is its interdisciplinary character and multidimensional scope, ranging from basic to applied knowledge and engineering, and including also an ontology and cultural narrative about life and living organisms and entities and, implicitly, about the depth of intervention of science in nature. Finally, the synthetic biology research community is not only building its cognitive and professional identity but has also, from the outset, devoted significant attention to ethical, risk and social analyses, with the declared goal of avoiding the mishaps encountered in the case of earlier biotechnology. It is for these reasons that this scientific area merits closer sociological analysis, despite what is still a low level of salience in the public opinion domain. In this chapter, we will offer a particular type of social science analysis, applying mainly the "public perceptions of science" framework and concentrating on just one subset of the issues raised by synthetic biology. Given the field's relative youth and its continuities with prior biotechnology, we will approach its treatment in an indirect way. We will start by examining for our purpose here the latest perspective of "public perceptions of science" and presenting some of its main results in the case of biotechnology (routinely used as a background case for the analysis of synthetic biology), before going on to delineate public views on synthetic biology and, from there, to extrapolate about how its applications may be received in the not so distant future.

## 6.2  The Space and Role of Public Perceptions of Emergent Technoscientific Areas

Since the turn of this century, public perceptions of science have been a relevant input to regulatory agencies, the scientific community, high tech companies and other stakeholders, particularly in Europe. This is a significant novelty. Until the last decades of the 20th century, public views on scientific advances were not part of the "public opinion" landscape and played no role in the regulatory process. That is not to say that, from time to time, some newly arrived scientific application did not spark a degree of controversy, which in most cases was fairly short-lived. But, generally speaking, until the last part of the 20th century, most scientific advances had a public profile that can be characterized as follows: (a) low public awareness or salience in comparison with other areas (politics, economic affairs, sports, entertainment), (b) only a small segment of the public regularly followed scientific news (the so-called "attentive public", that is, individuals interested and informed, albeit at an elementary level, about science; usually around 10 % of the population in most societies), (c) either they were positively received (the shared assumption being that virtually any scientific advance was good and an indicator of "progress") or else they were not problematic, (d) to benefit from them it was not necessary to understand the substantive content, but only to possess a fairly basic operational knowledge ("push buttons") and rely on the appropriate experts, and (e) there was a high level of trust in the scientific community and science-based technical expertise. Science and progress were perceived as strongly associated, reinforcing each other, and that link was a core component of the modern cultural narrative and the mindset of "modern" individuals.

However, from the late 1950s up to the present, a number of episodes of unease and even resistance to subsets of science and technology began to emerge; some of the most salient were nuclear armament, advanced automation (robots in industrial processes), fertilizers and pesticides (core elements of highly intensive agricultural practices supported by chemistry), nuclear energy, human embryo research, genetic modification of plants and animals, cloning, and, more recently, embryonic stem cell research. Resistance did not arise from a single overarching motive, but was associated with diverse factors, chief among them: (a) opposition to war and the arms race, particularly regarding nuclear and other weapons of mass destruction (represented by new pacifist movements), (b) the emergence in the mid-1960s of the so-called "environmental consciousness" and "green" values, (c) an explosion of publicly available information and media coverage on the risks linked, directly or indirectly, to science and technology, contributing to the promotion of a culture of benefits without side effects), (d) clashes between, on the one hand, ethical and religious creeds and, on the other, embryo research (clashes linked to conflicting views on the definition of when individual human life begins and the moral status and rights of an early human embryo), (e) the emergence of so-called "postmaterialist values" among the younger generations, and (f) the postmodernist critique of science and technology by influential figures in the humanities and the social sciences, not merely calling into question the side effects of science and technology, but also challenging the special epistemological status of scientific theories, and praising instead so-called "local knowledge" based on direct, intuitive and holistic experience (unaided by conceptual tools such as analysis, abstraction and formalization, or appropriately specialized instruments) open not just to the professional scientist.

The increase in mass media coverage of scientific and technological controversies has been a significant factor fueling public anxieties about certain scientific and technological advances. Media studies have shown that mass communication has a stronger influence on the "about what" than on the "what", and thus plays a major role in "agenda setting", giving salience to certain objects and issues while demoting or silencing many others. More specifically, according to the thesis of the "amount of coverage" of technological controversies, what really matters is not so much the balanced or unbalanced nature of coverage, but the sheer quantity of news devoted to potentially problematic areas, with a high coverage interpreted by the public as a signal that there is something wrong with a particular scientific development (see Mazur 1981).

The risks and ethical aspects associated with scientific developments and technological applications are the two dimensions that have received the most attention from new academic specialties. Risk analysis and bioethics have been the two main conceptual instruments deployed to address the regulatory challenges posed by disruptive technological innovations and, particularly, to address the latent concerns of a skeptical or potentially critical public regarding recent scientific advances. Most of the subsets of science that have come under open criticism in the last two decades are part of the life sciences, and this helps to explain the emergence and public prominence of bioethics as a field, and the proliferation of committees officially appointed to provide reports and advice to policy-makers on the ethical dimensions of new scientific developments.

While risk analysis research was quick to factor psychosocial and cultural variables into its models, such as the mental mechanisms employed by non-experts in the perception and ranking of technoscientific risks, usually at odds with the criteria used by experts (Slovic 2000), the large majority of bioethical analyses have paid little or no heed to any views or ethical judgments that the public might form, despite the availability of empirically grounded social science analyses of values and ethical views.

However, in the last few years a small but growing segment of the bioethical literature has taken as its starting point empirical data, including "public perceptions of science" (Lassen et al. 2006; Molewijk et al. 2004; Sugarman 2004). This recent approach still tends to operate under an implicit division of labor between bioethics and public perception studies, which consigns the latter to a very limited role; that of describing people's values and moral standards.[1] But this division of domains (bioethics dealing with the theoretical and normative ethical dimensions of new technoscientific developments versus empirical social studies confined to describing and offering "raw" data of values and perceptions) fails to take into account a number of relevant points.

Social science studies of values and perceptions of science not only describe ("measure") people's ethical mindset, but also analyze its formal properties such

[1]Felix Thiele's chapter "The Ethical Evaluation of Pharming", in Rehbinder et al. (2009), pp. 179–200, and the much earlier contribution by Birnbacher (1999) give more scope to social science studies in the framing of bioethical questions and objects.

as depth and structure (scope, consistency or inconsistency), its variability in the population and offer explanatory schemas on its underlying whys or "reasons". These studies, empirically grounded, show that moral values interact with other more general values and worldviews (such as views on nature and naturalness, demarcation between species, the role of humans in relation to nature, science and progress, religion, equality, peace, democracy) that are important for understanding more specific ethical evaluations. What from a bioethical perspective may seem to be inconsistent values held by an individual, from a social science angle may be understood as the result of individuals, for varied reasons, giving different weights to diverse frames[2] when evaluating a specific situation; distinct in most cases from those they would apply in another context. Social science analyses show that, when confronting a moral issue, most individuals, except principally those operating with a tightly integrated ideological structure—usually belonging to or identified with specific organizations with an explicit culture or credo—do not engage in a deductive chain of thought starting from a fundamental and overarching ethical or normative principle, but tend to apply a piecemeal approach, sensitive to the specifics and salient attributes of the issue at hand. Furthermore, moral views on central issues posed by the life sciences exhibit large variability within and between societies, and this is relevant for understanding the components and sources of moral concern and, obviously, for ethical debate and regulation in pluralistic societies. To sum up, studies of public perceptions of science are today not only descriptive but also explanatory. Although they usually lack an explicit normative dimension, in contrast with bioethics, they may have a role in the multidimensional evaluation of potentially controversial technoscientific developments. In turn, social science studies could benefit from the formal objects, concepts and explanatory schemas constructed by the bioethical and, more generally, the philosophical literature.

To recapitulate, while in the last two decades regulators sought to ground their decisions regarding scientific advances on analyses of risks and benefits and bioethics recommendations, more recently they have realized that here, as in other more classical domains, such as economic policy, open to public scrutiny and demands, public opinion cannot be ignored. In the case of the European Union, studies on public perceptions of science and technology have become a useful input for policy-makers, finding a place beside the standard soundings on economic or political issues. The interdisciplinary field known as "public perceptions of science" is of particular interest when public bodies are confronted with the challenge of balancing diverse goals in a democratic regulatory process, and it is also relevant for the scientific community and its interaction and communication with today's public.

---

[2]Frames are one of the two main components of "schema theories" in cognitive psychology and artificial intelligence (the other is "script"), and were introduced by computer scientist Marvin Minsky to denote the representation and the structure of a given piece of knowledge. Frames store an object in memory as a list of its most typical attributes, and when an individual hears, sees or reads something about the object, he/she "recalls" these attributes en bloc.

## 6.3 An Enlightenment Assumption: Resistance to Science as a Function of Lack of Knowledge

In the last part of the 20th century, diagnosis of the limited cases of opposition to science was reduced to a single factor: poor public understanding or, to give it the conventional label, low scientific literacy. This contemporary embodiment of an Enlightenment assumption—the attribution of opposition to material and institutional progress to a cognitive deficit or cultural barriers—rapidly gained currency among the scientific community, regulators and early public perceptions of science analysts. Accordingly, the recommendation or remedy was straightforward: the improvement of the formal (via education) and informal (via popularization) transmission of science to the public at large. Both, analysis and course of action were canonically represented by an influential report, prepared by the Royal Society in the UK under the title *The Public Understanding of Science* (The Royal Society of London 1985).[3] This approach has come to be labeled as the "deficit model".

Neither the scientific community nor public decision-makers paid any real heed to factors of resistance other than the low scientific literacy of the public. The public's values and preferences, the social and political culture and the worldview articulated by new social movements such as the "Greens" received either scant consideration or none at all. The skirmishes between the scientific community and the public were interpreted as a collision between, on the one hand, competence, rationality and progress and, on the other, public ignorance, unfounded or irrational fears and attitudes labeled "neo-Luddite" (in reference to the early resistance to and destruction of the new textile machinery by displaced artisans during the 19th-century Industrial Revolution in England).

The scientific community interpreted the critiques of specific subsets of science as a general frontal assault on two pillars of science as an institution: its funding with public monies, and its autonomy in setting scientific goals and conducting research. In this charged context, calls were made to promote public scientific literacy. The resulting program was deficient in its conceptual basis and patronizing in its orientation: it was assumed that anxieties and resistance to specific subsets of science and technology were without merit from a "rational" perspective and would be dissolved simply by improving the public's familiarity with the concepts and methods of science. All that was required of the scientific community was to leave their laboratories, not only to talk to policy-makers and entrepreneurs (the direct providers of financial resources), but also to devote time to explaining science to the public. However, this was a task with no immediate return and not likely to advance a scientist's career.

A decade later, a contrasting and more nuanced view of public resistance to science and how best to contend with it started to take shape, and was captured by the

---

[3]For a brief history and background of the report and the Public Understanding of Science field, as related by the scientist who chaired the ad hoc group appointed by the Royal Society, see Bodmer (2010).

House of Lords report *Science and Society* (House of Lords Select Committee on Science and Technology 2000; see also Einsiedel et al. 2006) The context of the science–public relationship was characterized then as a crisis of trust, and variables other than just scientific literacy were identified as responsible for the deficit. Improving the level of understanding of science was still perceived as relevant, but a new climate of openness, dialogue and "engagement" of the scientific community with the public was advocated.[4] New forms of public participation were also encouraged (for example deliberative polling, standing consultative panels, citizens' juries, consensus conferences, internet dialogues). In sum, the recipe this time was participation, dialogue, and a new science-society contract. Although the conceptual model of this new framework was not without its flaws, it was instrumental in fostering a conceptual and institutional space for legitimate differences in values and goals, and not just in levels of knowledge, between the professional scientist and the lay public. The net practical effects of the new mechanisms to give the public its say on potentially conflictive scientific areas have been subject to evaluation, but no clear evidence has been found that they have succeeded in overcoming public anxieties and reservations, although they may have contributed to offering new insights to decision-makers.[5] Despite this shortcoming, the new framework has been a significant step forward in enlarging the conventional view on public responses to science, indirectly stimulating a new generation of studies.

Recently, a more robust and general approach to explaining public views of science is in the process of being developed.[6] Its main novelty is to postulate that not only formal knowledge of science, but also culture count toward shaping evaluations of and attitudes to science. Some of the main elements and findings of this line of work are the following: (a) knowledge plays a role in the evaluation of science in general, with a positive correlation between knowledge and attitudes (more knowledge tends to be associated with more positive attitudes), but the size of its independent and direct contribution is small (Allum et al. 2008; Bauer et al. 2007; Pardo and Calvo 2006a), (b) unless there is a personal and strongly vested interest at stake, most people don't engage in a systematic, time consuming search for information in evaluating complex objects such as new scientific developments, but make use of a minimalistic and informal template composed of a straightforward hierarchy of goals and means (means-goals tradeoffs), that is, if the goals are highly valued (for example, potential gains in health) most individuals are willing to compromise regarding the means—assuming these don't directly impinge on core values (Pardo et al. 2009), (c) worldviews, values and specific cultural frames have a very significant influence in shaping views of some scientific and technological advances (Pardo and Calvo 2008), that is, attitudes to specific subsets of science,

---

[4]On the varieties of engagement, see Rowe and Frewer (2005).

[5]See methodological considerations about the evaluation of public participation forms in science policies in Rowe et al. (2004), Rowe and Frewer (2000, 2004), Levitt et al. (2005), and Nielsen et al. (2011).

[6]A sample of this work can be found in Bauer et al. (2012).

like biotechnology, are influenced by constructs embedded in the general culture of a society such as general expectations about science, belief in progress, images of nature and "naturalness" and associated beliefs about the role of humans in nature and "natural processes", views on animal and human life and the demarcation between the domains of the natural and of the artificial or designed, perceptions of non-human animals and their status in relation to humans, views on the moral status of human embryos, (d) risk perceptions play an independent role in shaping views on science developments, and (e) "trust" or *Vertrauen* in the scientific community and the regulator (based on perceptions of scientists' respect for the general interest) and "confidence" or *Zutrauen* (grounded on perceptions of technical competence, that is, the belief that researchers and experts, in their professional domain, know what they are doing better than anybody else) also have a significant influence on attitudes to science and may compensate for a modest level of scientific literacy (Priest et al. 2003; Siegrist 2000; Pardo 2012). In conclusion, according to current views from cognitive science, social psychology and cognitive and cultural sociology, people don't use mainly formal, systematic or "textbook" knowledge to make sense of complex issues such as scientific developments and their acceptability, but rely on an array of diverse ideational, symbolic and emotional elements present in the culture of a society at a given time (worldviews, beliefs, frames, values, fears and "taboos", reference groups, trust) and readily available for "activation" by most individuals as shortcuts, heuristics and "gut feelings" (including the so-called "yuck factor") when confronted with specific situations and decisions.[7] From this perspective, the until recently canonical approach of resorting to levels of scientific literacy alone to understand people's views of science would, in the best case, be no more than a limited subset of a more general explanatory framework that incorporates a much larger array of cultural components and captures not just the valence of attitudes (positive or negative), but their actual content.[8]

## 6.4 Changes in the Institutional and Cultural Framework for the Development of New Areas of Science

It is important not to exaggerate the scope and depth of current criticism of science. Contrary to postmodern characterizations of the status of science at the end of the 20th century,[9] it is safe to affirm that science exhibits a positive valence in today's culture and that this encompassing "environmental" influence corresponds

---

[7]For an early formulation of the "cognitive miser" model (the use of shortcuts and heuristics to overcome limitations in information processing), see Taylor (1981). On holistic and intuitive cognitive processing, see Gigerenzer (2007).

[8]On this point of characterizing the substantive content of evaluations and not only their sign (positive, negative), see Pardo and Calvo (2006a).

[9]For a review of this critique, see Heise (2004).

with favorable general perceptions of science by the public, as documented by numerous surveys on both sides of the Atlantic. The standard case of science-society interaction is represented by the following features. Firstly, most scientific developments take place silently, contributing to the continuous expansion of knowledge but known only to the corresponding subsets of the scientific community. They also tend to receive virtually no media attention, although this is compensated by a growing projection of research results through the Internet on websites maintained by universities, public research organizations, scientific academies, hospitals and industrial laboratories. Secondly, the vast majority of applications of science to satisfy social demands through technology are not problematic for the majority of people, and most of them are perceived as clearly beneficial. Thirdly, these technological advances embedded in new products or services become integrated into the background of the complex mode of the collective satisfaction of needs, without the public playing any direct role other than their more or less smooth, rapid or slow adoption. Fourthly, only very specific cases of scientific advances attain a high level of salience while still in the laboratory, due to their particular attributes and potential for fostering radical innovations that are socially, environmentally or culturally disruptive. Fifthly, cases of significant resistance to or open rejection of scientific developments by the public are an anomaly. These currently occur almost exclusively with certain subsets in the domain of one of the most dynamic scientific areas—that of biotechnology— whereas information technologies are viewed in a clearly positive light and eagerly embraced in a growing number of areas. At the current time, public awareness about another emergent area, that of nanotechnology, is medium to low, and developments are perceived for the most part as non-controversial (Cobb and Macoubrie 2004; Einsiedel 2005). Other technological areas that could be an object of controversy are developments that affect the natural environment, such as the technique for natural gas extraction known as "fracking".

Frequently, diagnoses of the reservations felt towards biotechnology in European societies are still based on the old template of a low level of scientific literacy and, in particular, the public's lack of familiarity with basic genetic concepts and principles. Certainly, there is ample evidence for this specific deficit of scientific literacy (in Europe, for instance, the much quoted, erroneous belief, documented by Eurobarometer surveys, that "genetically modified tomatoes contain genes while ordinary tomatoes do not"). But the literature consistently shows that this cognitive deficit is not sufficient to explain the complexity of public perceptions of biotechnology (Bauer and Gaskell 2002; Pardo et al. 2002). To make sense of that complexity, one would have to consider not solely knowledge of science but also general and specific variables, focusing on cultural and psychosocial factors and on the perceived salient attributes of biotechnology.

Two very general interrelated factors to take into account are that the institutional architecture for public decision-making and, also, the cultural framework for biotechnology research and applications today are significantly different from those surrounding the technoscientific areas which emerged earlier in the 20th century (including nuclear power in its initial years and almost up to the late

1960s). Institutional and cultural variables are associated and, thus, when a new scientific or technological development impinges on the "social mindscape" of a society,[10] impacting on the worldviews, frames (objects and their most salient properties) and core values of contemporary culture, the institutional or governance dimension of science may take on a significant importance, with mounting pressure for public participation or "voice".

From the mid-19th century, decision-making in technical or complex areas has been reserved for those individuals who, through a lengthy and demanding process of knowledge acquisition and parallel mechanisms of formal accreditation of the competencies thus attained, can exhibit the credentials of being legitimate members of a professional group, usually affiliated with a university department or research organization in a clearly demarcated field of knowledge.[11] In contrast, since the turn of this century we are witnessing growing pressure for the extension of the democratic principle to what was once considered the exclusive preserve of expert opinion, reserved for the professional scientist. The creation of a public entrance to the science domain, even if at present only limited access is granted, and this only in very special cases, marks a far-reaching institutional change, based on a principle neatly summed up in the dictum of French physicist and essayist Jean Marc Lévy-Leblond that "conscience should take precedence over competence", that is, no knowledge prerequisites should be imposed on individuals for expressing their views and preferences as a vote (Lévy-Leblond 1992). If, until recently, the view of Lévy-Leblond was very much a minority one with no real practical repercussions, today it is in the process of becoming an assumption that is taken for granted in a number of societies. And although the science domain is one of the few areas in which a large percentage of the public is willing to leave it to the experts (scientists) to make technical decisions that affect society at large, the canonical combination of external oversight by bureaucratic agencies and the principle of "self-regulation" by the scientific community is no longer perceived as sufficient for a healthy governance of science. The dramatic decline of trust in most public institutions and, secondarily, elite (professional) decision-making observed in many advanced societies since the last part of the 20th century is contributing to the creation of new spaces for public participation (see Norris 1999; Nye et al. 1997). Not only the social scientist but also the regulator knows that to give "voice" to the public on scientific policies in sensitive areas (such as biotechnology), or at least to take into account its views as gauged by surveys and other means, may be critical in avoiding the alternative and more costly course of action: namely the public's "exit" or alienation from and opposition to new scientific developments, to borrow the elegant typology of Albert Hirschman, proposed for a different domain (Hirschman 1970).

---

[10]On the concept of "social mindscape" see Zerubavel (1997).

[11]On the emergence of the role of scientist and the process of institutionalization of science, see Ben-David (1984).

In parallel with the novelties in the institutional landscape in which science operates, an even more far-reaching change has taken place in the cultural domain, deeply affecting the social reception of the life sciences and, especially, of bio-technology. Paradoxically, this novelty is due in part to the very development of science, particularly of ecology and conservation biology: new academic disciplines that have not only provided solid theoretical models and robust empirical evidence on the sorry state of the natural environment and its anthropogenic origin, but have also issued calls for action, influenced general ideas and narratives about nature (such as the ones portrayed by literature and films), and, indirectly, contributed to shaping values, feelings and emotions about its conservation. This is not to say that ecology as a scientific field and environmental consciousness and green social movements share the same conceptual basis. Each of them has its specific roots, traditions and conceptual schemas, some of them at odds with each other: specifically, the analytical and to some extent value-free approach of ecology versus the critique or skepticism about the scientific approach as a way of acquiring knowledge and its detrimental practical effects embraced by subsets of the environmental movement (see Bowler and Morus 2005, pp. 213–236; Marx 1988, pp. 160–178; Pepper 1996, pp. 239–294). In contrast to what happens in other areas of science, research in ecology and, particularly, conservation biology is guided by external goals or has an explicit commitment to conserve and restore "natural" ecosystems and species, and it is not uncommon to hear distinguished researchers advocating for public action to reverse the conventional course of society-natural environment interactions. New scientific knowledge in the environmental sciences (recently on climate change) and any associated policy proposals are given salience by the media, and environmental organizations translate them to calls for individual and collective action. Also, in the last two decades, in addition to compelling narratives about nature offered by fictional works in literature and films, there has been an explosion of vivid nature documentaries, many of them based on scientific evidence, offered by TV channels and available on the Internet and other electronic media, which are contributing to shape and activate an environmentally friendly "mindscape". The roles of the scientific community, the media and environmental associations, coupled with people's personal, macroscopic experience with the environmental side effects of the dominant model of growth (supported by intensive use of science-based technologies and inattentive until very recently to environmental negative externalities), have eroded the belief in progress and, above all, the assumption that any possibility for action, opened up by scientific advances, has a positive valence and should be supported.[12] We are witnessing the emergence of motifs and elements of a neo-Romantic, pastoral vision of nature, dramatically different from the Enlightenment-rooted perspective that accompanied modernity and industrial society until the mid-1960s of the 20th century. This reborn Romantic vision has among its main properties being holistic,

---

[12]Cultural historian Leo Marx has attributed the loss of collective optimism and the erosion in the narrative of progress to awareness of "the grave damage that modern industrial societies inflict upon the global environment" (see Marx 2001).

based on feelings, aestheticism and an aspiration to heal the wounds humans have inflicted on nature through industrial production and mass consumption sustained by science and technology.

Erosion of the idea of progress does not signify abandonment and replacement of one of the most powerful narratives of modernity, still active in the second decade of the 21st century.[13] But a large part of the population in many advanced societies, while believing in progress and the beneficial role of science, is more alert now to the side effects and advocates for a more restrained intervention of science and technology in the "natural domain". And this worldview in flux has already had deep implications for specific areas of science such as biotechnology, perceived by many as a radical, unacceptable way of crossing natural boundaries and altering the very last frontier of "naturalness". It also seems likely that it could influence the reception of some synthetic biology applications, once they become available and reach a sufficient level of public awareness (which, as we will see later, is now virtually absent). For this reason, it is of paramount importance to grasp the main elements of the vision of nature embedded in the culture of the present decade.

The current components of the worldview of nature and the role of science within it are documented in a recent international study on values and worldviews, based on a large survey applied to representative samples of the adult population in ten European countries, Russia, Japan, the United States and Mexico.[14] The main results relevant for our purpose are set out below.

An almost universal view in the 14 countries analyzed is that nature has a systemic character (that is, it is a set of interconnected elements), and is both extremely fragile and acutely sensitive to human activities. The view that nature would be at peace and in equilibrium if left untouched by humans also tends to predominate, despite meeting with less consensus in Japan, Denmark, the Netherlands, the USA and Sweden (see Table 6.1).

A second vector of the current worldview refers to the aesthetic and emotional attachment of humans to nature. In all the societies considered, the aesthetic dimension of nature is ranked above the artificial or "built environment" ("the things built or created by humans"), albeit slightly less pronouncedly in Denmark, the Netherlands, France and Germany. A feeling of "peace and tranquility" is

---

[13]For the case of Germany, characterized both by a high appreciation of science and also of nature and environmental values, see Institut für Demoskopie Allensbach, *Kein Fortschrittspessimismus. Eine Dokumentation des Beitrags von Dr. Thomas Petersen in der Frankfurter Allgemeine Zeitung* Nr. 115 vom 18. Mai 2011, available at http://www.ifd-allensbach.de/uploads/tx_reportsndocs/Mai11_Fortschritt.pdf, accessed 22 July 2015.

[14]Information was gathered through a survey of 21,000 people aged 18 and over in 10 European Union countries (Sweden, Denmark, the United Kingdom, Germany, Netherlands, France, Italy, Spain, the Czech Republic and Poland), Russia, Japan, the USA and Mexico. Fieldwork was conducted by Ipsos and completed in January 2013. Sample size of 1,500 cases in each country. The design and analysis of the study are the work of R. Pardo, M. Szmulewicz, J. Maquet and C. Perera at the BBVA Foundation Department of Social Studies and Public Opinion.

**Table 6.1**  Statements reflecting a systemic and fragile view of nature

| | Nature is a set of interconnected elements | Nature is very fragile | Nature's balance is extremely delicate and can easily be altered by human activities | Nature would be in peace and harmony if human beings would leave it alone |
|---|---|---|---|---|
| Total EU countries (10) | 7.9 | 7.6 | 7.5 | 7.3 |
| Sweden | 8.5 | 7.9 | 7.5 | 6.6 |
| Denmark | 7.8 | 7.0 | 7.1 | 6.3 |
| UK | 7.5 | 7.7 | 7.5 | 7.0 |
| Germany | 8.1 | 7.4 | 7.5 | 7.3 |
| Netherlands | 7.7 | 7.5 | 7.3 | 6.3 |
| France | 7.9 | 8.2 | 8.0 | 7.3 |
| Italy | 8.2 | 7.4 | 7.6 | 7.8 |
| Spain | 7.8 | 7.7 | 7.5 | 7.5 |
| Czech Rep. | 8.2 | 8.0 | 7.7 | 7.3 |
| Poland | 7.3 | 6.9 | 7.0 | 7.0 |
| Russia | 8.4 | 8.4 | 7.9 | 7.7 |
| Japan | 7.6 | 7.9 | 7.3 | 5.6 |
| USA | 7.1 | 7.2 | 6.8 | 6.4 |
| Mexico | 6.8 | 7.4 | 6.9 | 7.0 |

"I would like you to tell me how much you agree or disagree with the statements I am going to read out." Means on a scale from 0 to 10 where 0 means "totally disagree" and 10 "totally agree". Base: all cases

associated with nature in the fourteen societies, with very high mean values (above 7.5 on a scale from 0 to 10) (see Table 6.2).

A third component of this worldview refers to the presence or absence of the canonical materialist or utilitarian view of nature that goes back at least to the Enlightenment, and that was part of the narratives of modernity and industrialization (see Table 6.3).[15] The strongest formulation of this vision is captured by the statement that "nature exists to be dominated by human beings", which is generally rejected, but with variability within and between countries: strong rejection in Denmark, Sweden and Japan, followed at a distance by France, Spain and Germany; less pronounced rejection in the USA, Mexico, the Czech Republic, the Netherlands, Italy, Russia, and the United Kingdom; and, finally, approval in Poland alone. Europeans are divided, both between and within countries, as to whether "the exploitation of nature is unavoidable if humanity is to progress" (mean value of 5.2), but in general this less radical facet meets with wider acceptance than the idea of humans dominating nature. The exceptions are Germany and

---

[15]This vision is already present in the Bible (Genesis chapter 1–28.29) and has been identified as a cultural component of the contemporary ecological crisis; see L. White, Jr' influential paper "The historical roots of our ecological crisis" (1967).

**Table 6.2** Statements reflecting an aesthetic and emotional attachment of humans to nature

| | Natural things are more beautiful than the things built or created by human beings | Nature gives me feelings of peace and tranquility |
|---|---|---|
| Total EU countries (10) | 7.3 | 8.1 |
| Sweden | 7.0 | 8.9 |
| Denmark | 6.6 | 8.6 |
| UK | 7.6 | 7.7 |
| Germany | 6.9 | 7.9 |
| Netherlands | 6.6 | 7.9 |
| France | 6.7 | 8.3 |
| Italy | 7.9 | 8.4 |
| Spain | 7.8 | 8.5 |
| Czech Republic | 7.5 | 8.1 |
| Poland | 7.0 | 7.5 |
| Russia | 7.6 | 8.3 |
| Japan | 7.8 | 8.1 |
| USA | 7.4 | 7.8 |
| Mexico | 7.5 | 7.7 |

"I would like you to tell me how much you agree or disagree with the statements I am going to read out." Means on a scale from 0 to 10 where 0 means "totally disagree" and 10 "totally agree". Base: all cases

**Table 6.3** Statements reflecting an instrumental view of nature

| | Nature exists to be dominated by human beings | The exploitation of nature is unavoidable if humanity is to progress | Economic growth is more important than protecting the environment |
|---|---|---|---|
| Total EU countries (10) | 3.9 | 5.2 | 4.3 |
| Czech Rep. | 4.5 | 6.9 | 4.4 |
| Poland | 5.7 | 6.5 | 5.5 |
| Germany | 3.7 | 4.4 | 3.7 |
| Netherlands | 4.4 | 4.4 | 4.3 |
| UK | 4.2 | 5.7 | 4.7 |
| Spain | 3.5 | 4.7 | 4.2 |
| France | 3.1 | 5.7 | 4.2 |
| Denmark | 2.4 | 6.2 | 3.0 |
| Sweden | 2.6 | 5.1 | 2.9 |
| Italy | 4.3 | 5.1 | 4.8 |
| Russia | 4.3 | 6.9 | 4.3 |
| Japan | 2.9 | 4.5 | 4.3 |
| USA | 4.6 | 5.0 | 5.0 |
| Mexico | 4.6 | 5.1 | 4.9 |

"I would like you to tell me tell me how much you agree or disagree with each of the statements I am going to read out." Means on a scale from 0 to 10 where 0 means "totally disagree" and 10 "totally agree". Base: all cases

the Netherlands, where this view of the inevitability of the exploitation of nature is rejected by the majority. Finally, the critical tradeoff of economic growth versus the protection of the natural environment, favoring the former at the latter's expense (a finding that is routinely taken as an indicator of the persistence of materialist values), is rejected by the population in all countries except Poland and the USA. In three societies (Mexico, Italy and the United Kingdom) there is a sharp division between the two views. The strongest rejection is observed in Sweden, Denmark and Germany.

As remarked before, the environmental mindset has a rather uneasy relationship with science. On the one hand, it draws on scientific knowledge about the state of the environment through fields like ecology, conservation biology and the interdisciplinary field of climate change. On the other, there is a widespread perception, particularly strong among members of environmental organizations, that science and technology, coupled with a growth model uncaring about negative externalities, are at the root of the current destruction of ecosystems and species extinction.[16] Scientific and technological progress can be seen then as part of the solution or, alternatively, as a threat, and most people tend to perceive the dual character of science in this respect.

Most citizens believe that science provides a potent tool with which to understand nature (mean value of 7.2 in the ten European countries and slightly lower in all remaining societies) and also that science can help to solve major environmental problems such as climate change. But at the same time, they see clear boundaries in the natural domain beyond which science must not trespass: in the 14 societies a clear majority believes that "scientists should not interfere with or change the workings of nature" (mean value of 6.7 in the ten European countries and Russia, and slightly below in all other countries). This position is connected with the very strong view that "any change man makes in nature will be for the worse, even if it is grounded in science", which is approved in all but three (the two Scandinavian countries and the Netherlands) of the fourteen societies surveyed (see Table 6.4).

Finally, a much debated reason for pro-environmental attitudes is the belief that "human beings should respect nature because it is the work of God". This idea has a clear association with the level of religiosity in each society and, accordingly, support is particularly high in countries characterized by a high degree of religiosity: Russia (mean of 7.4), the USA (mean of 7.0), Poland (mean of 6.9), Mexico (mean of 6.8) and Italy (mean of 6.5) (see Table 6.5). But the fact that this idea finds majority support in 11 of the 14 societies considered, even in countries like France with a strong secular culture, shows that it has found its way into the contemporary mindscape or culture at large, decoupled from or deflating the literal religious interpretation.

---

[16]For a general picture of the historical link between science development (including ecology) and the accompanying narrative and the exploitation of new natural resources and areas of the planet, see Bowler and Morus (2005), pp. 213–236.

**Table 6.4** Statements on the role of science regarding nature

| | Thanks to science we now have a better understanding of how nature works | Science and technology have a vital role to play in solving the problem of Earth's global warming | Scientists should not interfere with or change the workings of nature | Any change man makes in nature will be for the worse, even if it is grounded on science |
|---|---|---|---|---|
| Total EU countries (10) | 7.2 | 6.8 | 6.7 | 5.9 |
| Sweden | 7.9 | 7.2 | 6.3 | 4.1 |
| Denmark | 7.9 | 7.4 | 5.5 | 3.9 |
| UK | 7.6 | 7.0 | 6.5 | 5.5 |
| Germany | 6.7 | 6.5 | 6.8 | 6.3 |
| Netherlands | 7.3 | 6.4 | 5.8 | 4.5 |
| France | 7.7 | 7.0 | 6.8 | 5.7 |
| Italy | 7.1 | 6.8 | 7.3 | 6.3 |
| Spain | 7.4 | 6.9 | 6.6 | 6.1 |
| Czech Republic | 6.7 | 6.3 | 6.6 | 6.2 |
| Poland | 7.0 | 6.5 | 6.5 | 6.1 |
| Russia | 7.3 | 6.6 | 6.7 | 7.6 |
| Japan | 6.1 | 6.6 | 6.4 | 5.5 |
| USA | 6.7 | 6.4 | 6.3 | 5.5 |
| Mexico | 6.6 | 6.5 | 6.3 | 6.3 |

"I would like you to tell me how much you agree or disagree with each of the statements I am going to read out". Means on a scale from 0 to 10 where 0 means "totally disagree" and 10 "totally agree". Base: all cases

**Table 6.5** Respect for nature because it is the result of God's work

| Human beings should respect nature because it is the work of God | |
|---|---|
| Total EU countries (10) | 5.2 |
| Sweden | 3.5 |
| Denmark | 2.8 |
| UK | 5.7 |
| Germany | 6.2 |
| Netherlands | 4.7 |
| France | 5.0 |
| Italy | 6.5 |
| Spain | 5.2 |
| Czech Republic | 5.7 |
| Poland | 6.9 |
| Russia | 7.4 |
| Japan | 5.9 |
| USA | 7.0 |
| Mexico | 6.8 |

"I would like you to tell me how much you agree or disagree with each of the statements I am going to read out". Means on a scale from 0 to 10 where 0 means "totally disagree" and 10 "totally agree". Base: all cases

These data suggest that any new scientific development that could be perceived as compromising nature and radically departing from "naturalness" would probably activate the just-charted multifaceted worldview of nature, which today is part of the "mindset" of a large part of the population in advanced societies, thus triggering opposition and resistance to it.

Before we present the scant empirical evidence and review the main findings of the literature on the social reception of synthetic biology, it may be useful to briefly characterize the main profile of the public's response to biotechnology, since the new developments have some commonalities and perceived similarities with it, and because it is a matter of debate whether synthetic biology applications would trigger similar responses to those encountered in biotechnology. Also, the question of the existence in parallel of a culture positive about science (measured through positive attitudes to science at large) and opposed to just a few of its subsets has been analyzed with regard to biotechnology, and the findings may also be relevant for synthetic biology. In this context, it is also of interest to see whether it is more fruitful to focus on attitudes to an entire scientific or technological field (biotechnology, synthetic biology) than to specific applications with particular attributes.[17]

---

[17]In chapter two of this book it is argued that, in the case of synthetic biology, these specific attributes go beyond genetic engineering. Attributes include the depth of intervention that leads to a new dimension of uncertainty, the orthogonality that causes a stepwise genetic separation of synthetic organisms from natural organisms and the claim to create life with its ontological impact.

## 6.5 Positive Attitudes to Science and Reservations About Biotechnology

An analytical task needing more attention in the field devoted to the study of scientific culture is how to account for the simultaneous presence of clear positive predispositions to science at large and strong reservations to specific subsets or areas. The mere existence at a given time of diverse science domains (from astrophysics to biotechnology let's say) perceived to have opposite valences (positive to negative, through neutral) suggests that a number of salient traits specific to each domain trigger or are associated with different evaluative responses. And this, in turn, indicates that factors (cultural, socio-psychological, and institutional) other than knowledge are at play. Yet during at least its first decade of existence, the public understanding of science field worked on the implicit assumption that science was an undifferentiated "whole", perceived as such by the public. The field's main object of interest was general attitudes to science. And these attitudes or perceptions were measured through a haphazardly composed list of items and effects of science, without giving adequate weights to different, representative areas of science and technology, such that the inclusion in surveys of more or fewer items capturing potentially positive or, at the other extreme, problematic areas could have the effect of raising or lowering the global score.[18]

Some authors began to point out in the early 1990s that general attitudes to science have little predictive power for evaluations of markedly different scientific areas. Science as a whole might have a positive image in the minds of the public, but this is nonetheless compatible with critical attitudes to specific scientific areas. The best course, they argued, would be to focus on public perceptions of clusters of science and technology sharing properties or attributes that separated them from the rest (see Daamen et al. 1990). Other researchers took a different route, and instead of using a single score (resulting from adding items with opposite valences, positive and negative) to capture attitudes to science at large, or even to some of its subsets (such as biotechnology), advocated for distinguishing between two facets of science (or even two facets of a specific area of science), one with a positive valence (labeled "promise of science"), the other with a negative valence (labeled "reservations about science"), opening up the characterization of these two components and their interaction in different societies and for different areas of science (from virtually no correlation between them to a negative correlation of different magnitude) (Miller et al. 1997; Miller and Pardo 2000).

The research conducted in the last two decades has shown that even clusters of scientific areas and two evaluative facets aggregate too much. There are commonalities in public perceptions of a cluster such as biotechnology (perceptions in many countries tend to be more negative than positive), but people also

---

[18]For example, the inclusion in a battery of 10 items measuring general attitudes to science of just one item or statement capturing the effects of science and technology on armaments could lower the global score of positive attitudes (see Pardo and Calvo 2002).

discriminate by type of application/purpose (agriculture and food or "green bio-tech", biomedical or "red biotech") and type of means (roughly, plants, animals, humans). Even when there is an open or latent general distrust of the means employed—the modification through genetic engineering of the blueprint of plant and animal life, and even more so that of humans—in general the so-called "red" biotechnologies, of a biomedical nature, are favorably perceived or, at least, a num-ber of them do not meet with significant reservations (with the exception of cases that involve early human embryos or cloning), whereas "green" biotechnologies, focusing on the genetic modification of plants for agriculture and the production of foods (not for pharmaceuticals), are critically perceived (Gaskell et al. 2010; Gaskell et al. 2006; Pardo et al. 2002; Pardo and Calvo 2006b; Sjöberg 2004).

The application of the current conceptual schema allows us to see that the more or less active resistance to biotechnology of the first half of the 1990s has given way, in the early 21st century, to a more moderate opposition and even a positive evaluation of some applications and, more importantly, to a more flexible perspec-tive that discriminates according to the specifics both of the goals of the research and the means being utilized (Pardo et al. 2009). This new mindset may be simply an effect of increased familiarity with biotechnological applications, linked to the passage of time and the incremental and piecemeal societal integration of the cor-responding research applications, which gradually leave the foreground and recede into the background in a trajectory observed in other, earlier cases of technologies surrounded by controversy.

Critical views of biotechnology are not just a function of lack of biologi-cal knowledge (although this is very low and contributes to them). Two types of explanatory variables should thus be considered. Firstly, there are a number of *specific* evaluative angles which have been shown in the literature to be relevant to the perception of biotechnological developments (variables with a short distance to the object of the attitudes, whether biotechnology as a whole or specific biotech-nological applications), among them: "literacy on genetics", and, for each applica-tion, evaluation of its "usefulness" (perceptions of benefits and their distribution), "morality", "recklessness", "naturalness", "riskiness", "playing God", product of "scientists' arrogance or hubris", compromising the principle of "human dignity" and, in the case of animal biotechnology, incompatibility with the "dignity of ani-mals", "suffering of animals" and the question of their "rights". Secondly, there are more *general* variables (with a long distance to the attitudinal object), i.e., var-iables not immediately connected with a particular scientific development, known to influence people's views of particular areas of science (such as biotechnology): "scientific knowledge" (general scientific literacy), "general expectations about science" (positive and negative facets of attitudes to science as a whole), "images of nature" (instrumental, romantic), different facets of the "vision of animals" (perceived position on the sociozoological scale, perceived intelligence, perceived sentience), "general risk perceptions or attitudes", the pair "trust-confidence" in the scientific community and regulators, the "amount" and "style" (congenial, neu-tral, adversarial; balanced/unbalanced) of media coverage of science and technol-ogy, and the political and civic culture regarding "public participation" in policies in novel areas (for example, environment, health, food, science and technology). Culture has many strata and usually the variables closer to or sharing more specificity with the attitudinal target have more direct explanatory power than more distant or contextual variables, although the latter are still influential (see Pardo and Calvo 2008; Pardo and Calvo 2006a, particularly p. 10). In sum, a vast domain of elements, many of them pertaining to the cultural framework of a given society and period, shape the public's mindset and attitudes to specific scientific developments, particularly in the symbolically rich areas of the life and environmental sciences.

Scientific and technological developments, taken as a whole, play a very significant cultural role, shaping, enlarging and changing the frontiers of the "mindscape" of a given period and society, affecting even the perceived ontology (the population of objects and entities, natural and artificial, their status, autonomy, salient attributes, demarcation and interrelatedness), the language, the ideas, the beliefs, the values, the narrative, in sum, the cognitive schemas that individuals apply to interpret and evaluate the world, and to organize the realm of everyday experience.[19] In turn, culture at large interacts in many ways with the development of science: from prizing certain "themata" and presuppositions to demoting or ignoring many others;[20] from being more congenial to more critical, through more cautious or skeptical; from favoring a more theoretical, abstract orientation to a more applied and useful type of scientific research; from being more sensitive to the natural environment to favoring its artificial re-creation and radical modification for practical purposes, as the history of science and culture has documented.[21] Cultural metaphors also influence at least the first development stages of symbolically rich scientific fields such as biology (see Keller 1995, XIII–XVIII). Elites of individuals (intellectuals, artists, writers, influential and outspoken scientists and technologists, science communicators) and organizations (the mass media, policy-making institutions and interest groups with a stake in science) play a significant part in the cultural integration of science in a particular society or period.

---

[19]Gerald Holton has noted that "science has always had (…) a metaphoric function—that is, it generates an important part of a whole culture's symbolic vocabulary and provides some of the metaphysical bases and philosophical orientations of our ideology. (…) Ideas emerging from science are, and will continue to be as they have been since the seventeenth century, a central part of modern culture—through pure thought, through practical power, and through metaphoric influence" (Holton 1995, p. 129).

[20]One particularly influential preconception is "the generally accepted thema of the unlimited possibility of *doing* science, the belief that nature is, in principle, fully knowable" (Holton 1988, p. 18). Analysis and synthesis are also cultural presuppositions: "High on the list of achievements our culture has traditionally defined as best are grand, synoptic, and unifying works usually characterized as 'syntheses' of the thinking of a period or a field" (Holton 1978, p. 111).

[21]Science historian Thomas P. Hugues has characterized the relationship between science and technology with American society during the period 1870-1970 as a "century of invention and technological enthusiasm", shared both by the scientific community and technologists, entrepreneurs, the government and society at large (Hugues 2004).

This process, labeled "intellectual appropriation" of science and technology (Hård and Jamison 1988), takes place today at the local, national and global levels. National society, despite having fuzzier contours and less symbolic influence in the 2010s than three or four decades ago,[22] still operates as a culturally differentiated domain, and is accordingly an obligatory unit of analysis for explaining the variability in the interplay between science and culture at large. This cultural role of science and the more general culture in which is embedded, affecting its development, is an important axis for charting the trajectory of acceptance and integration of scientific areas with a disruptive potential; biotechnology and synthetic biology chief among them.

## 6.6  Public Views on an Emergent and Fuzzy Object: Synthetic Biology

Taking into account the immense domain that is science and technology today, where can we locate the relative visibility attained by synthetic biology in the public opinion mindscape? How can we calibrate the relative weight of risks, benefits, ethical issues and worldviews (particularly of "unnaturalness" and general attitudes to science) in the public's acceptance of synthetic biology applications? Would public views be equivalent to the by now familiar perceptions of biotechnology?[23] And might we therefore expect a trajectory of resistance followed by partial acceptance, similar to that encountered by biotechnology? Is it likely that demands will emerge for a public "voice" in synthetic biology developments, or will the public consent to a regulatory institutional architecture in which scientists, public officers and policy-makers shape the direction of the field and its translation into practical applications? These are the main questions addressed in the closing part of this chapter.

---

[22]Probably the following characterization is no longer valid, but a more relaxed one is still at work for many areas of public opinion and culture: "We tend to think of the 'country'—the particular nation-state we live in—as the maximal social unit not only of economic and political life, but also of social organization and culture, the 'way of life' we are part of. The nation-state has such special importance that many of us rarely think beyond it [...]", (Worsley 1987, p. 50).

[23]Biotechnology as a frame of reference for the analysis of the many dimensions of synthetic biology is a constant in the literature. For a defense of the differences between them regarding the ethical component, see Boldt and Müller (2008). For these authors, the ethical novelty of synthetic biology is the result of the fundamental discontinuity entailed by the transition from "manipulatio" to "creatio": "In synthetic biology, the aim is not to amend an organism with a certain quantity of altered characteristics (that is, to manipulate); instead, it is to equip a completely unqualified organism with a new quality of being (that is, to create a new form of life). [...] Seen from the perspective of synthetic biology, nature is a blank space to be filled with whatever we wish." (p. 388).

## 6.6.1 Low Level of Salience and the Study of Public Views

For an emergent scientific field to become institutionalized, it must first acquire critical mass, attract young researchers, have its early theoretical, methodological or practical novelties recognized by fellow scientists from the matrix discipline and neighbouring ones as making a distinctive and significant contribution and, last but not least, gain access to specific funding and academic positions. Synthetic biology is, in these respects, at an early stage, particularly in Europe.[24]

Even when a field has reached a certain level of institutionalization, the path to achieving salience beyond the laboratory, academic department or specialist meeting is not an easy one. Today, more than in the not so distant past, a level of public salience and a positive climate for its social reception may have an indirect but significant influence in the field's take-off and possibly even its maturity stage. The opportunity cost of supporting one specific area versus any other(s) is too high to be ignored by policy-makers, private investors and researchers.

Synthetic biology, moreover, faces two major barriers in the way of becoming an object of public perceptions.[25] One is its technical complexity, which poses considerable difficulties for the media and members of the public with little understanding of biology and genetics. In this sense, even a modest increase in the relevant scientific knowledge of journalists and the public may be helpful, and a clearly guided effort on the part of the scientific community to convey early on its key concepts and messages could make a difference.[26] The other barrier, connected with the first, is reaching a salience threshold (in terms of both quantity and recurrence) in the media. Salience will usually result from dramatic theoretical advances, the announcement of particularly important benefits (actually delivered, but also anticipated or hyped), powerful or highly symbolic narratives about the field, or, at the other extreme, mishaps or accidents, publicly aired disagreements among the experts, or campaigning in favor or against by influential stakeholders and organizations.

Despite the emphatically applied nature of synthetic biology, it has not so far produced substantial practical developments in domains such as human health, the economy, energy production or cleaning the environment. What there is, as

---

[24]For a study of national differences in public funding in Europe of both synthetic biology and ELSI (Ethical, Legal, Social Issues) linked to synthetic biology, see Pei et al. (2011).

[25]Although references are made in these pages to attitudes and views on synthetic biology as a whole, this is just shorthand for attitudes to specific synthetic biology applications. It is highly unlikely that most individuals will form or hold attitudes towards a new scientific field as such, unless the field becomes linked and identified with an overarching and public narrative (such as "regenesis", "creating life"). At the time of this writing, the expression "synthetic biology" for the large majority of people is not associated with strong traits, attributes or images, in contrast with genetic engineering and, particularly, of cloning.

[26]For a number of interesting findings apropos of the communication of synthetic biology, see Kronberger et al. (2009).

generally occurs with the advent of a new technoscientific field with a potentially strong applied focus, is an abundance of promises and hype, aimed more at building a professional identity, giving direction to the field and capturing the attention of policy-makers and potential financial backers than at gaining the interest of the general public. Still, leading researchers working in synthetic biology know quite well that in the current competitive market for resources, the attention of the media and the public could have a significant impact on the level of support obtained. And perhaps even more important, they are fully aware, due to the painful experience with biotechnology, that the valence (positive, negative, ambivalent) of early social reception may have an impact on the regulatory framework. One of the challenges, precisely, is to pitch their field against a myriad of other scientific domains and subfields, and particularly to demarcate it from conventional biotechnology. The absence at present of highly visible positive and/or adverse effects, the cautious approach of researchers in this area, who from the outset have shown themselves to be sensitive to the ethical and risk dimensions in order to avoid the problems besetting the biotechnology field, and the lack, at this time, of significant campaigning by organized groups explain the low public salience of synthetic biology. According to data from a European Commission Eurobarometer, conducted in 2010, 83 % of Europeans knew nothing or had never heard of synthetic biology (European Commission 2010; Gaskell et al. 2010). Given this low level of awareness, organizations (such as environmental groups) who might wish to make a public issue of synthetic biology developments have at present little room for manoeuvre. It could well happen that the field develops silently in the next few years, without major public attention and, correspondingly, no opposition, even though it may impact more deeply on the very dimensions that fueled rejection in the case of biotechnology applications. Today's situation regarding the social reception of synthetic biology has been aptly characterized in a recent paper under the expressive title "Looking for conflict and finding none" (Kaiser 2012), which contrasts the flurry of reports and special working groups on ethical, legal and social issues (ELSI) of synthetic biology with its virtual absence from the public's mindset, and calls into question central assumptions and even the meaning of early analyses of public perceptions of this area.[27] Its author, Matthias Kaiser, summarizing the main conclusions of a special issue of the journal *Public Understanding of Science* devoted to synthetic biology, has noted that, in Europe, general attitudes to science are positive; the scientific community is viewed favourably by the public; media coverage of synthetic biology has been ambivalent, "between fascination and repulsion", but without giving special weight to critical aspects of the associated developments (Gschmeidler and Seiringer 2012); and, finally, there are no clear signals that developments in synthetic biology would mimic the controversies over genetically modified foods (Torgersen and Hampel 2012). Kaiser also echoes the doubts and even skepticism voiced by a new generation of researchers

---

[27]For a review of ELSI reports on synthetic biology, see Torgersen (2009).

regarding the model of "public engagement", but advocates good governance mechanisms that take into account ELSI and, perhaps more significantly, people's latent values that could be affected or activated by new developments in this scientific field.

Although the currently underdeveloped status and low public awareness of the synthetic biology field limit the kind of public perceptions studies that can be conducted, it does not completely rule them out. However, the analysis route must be somewhat indirect and of an exploratory nature. Taking into account the dominant worldview of nature, general perceptions of science, the public's views on the role of scientists regarding the radical modification of nature (the "playing God" argument), the specific factors sustaining reservations towards biotechnological applications, the main attributes of the synthetic biology field and the early data available regarding its perception, it is possible to anticipate a number of plausible evaluative angles that the public will apply when the field reaches a threshold of maturity and coverage in the media. And, as has been the experience with biotechnology, public views can be expected to exhibit significant variability as a function of the specifics (that is, the goals and means) of the various applications mooted. It is also plausible that, even in the absence of high profile applications, a global positioning on synthetic biology could develop if leading researchers in the field insist on offering a highly symbolic narrative (perhaps necessary for demarcating the field from neighboring disciplines) that stands at odds with current worldviews and values: a narrative that portrays synthetic biology as "the *creation* of synthetic life in the laboratory", "the dawn of a new era in which new *life* is *made* to benefit humanity", as "a second *genesis*" or "*regenesis*",[28] as the production of "*living machines*"; a self-presentation aligned with characterizations of the field such as "extreme genetic engineering" offered by social analysts of science and technology.[29] And since synthetic biology touches on a highly sensitive domain (nature) and deeply rooted cultural dimensions (views of life, the role of humans regarding life and, particularly, the role of scientists in its modification or creation), we should expect significant national disparities deriving from cultural differences.

---

[28]Craig Venter quoted in The Guardian on Thursday 20 May 2010, under the title "Craig Venter creates synthetic life form". See also, Church and Regis, *Regenesis* (2012).

[29]See ETC Group (2007). Joachim Boldt and Oliver Müller have alerted synthetic biology researchers to the risks of using metaphors of the type mentioned that change the notion of life and blur the boundaries between organisms and artifacts (see Boldt and Müller 2008, p. 388), activating images of Faust or Frankenstein. See also the complementary letter to the editor under the title "Of Newtons and heretics" by Ganguli-Mitra et al. (2009), which presents the results of a survey conducted by its authors among 20 European synthetic biology researchers, showing that most of them had a narrow and traditional vision of the ethical problems posed by the field (mainly, issues of biosafety and biosecurity), virtually none of them connected to its larger purpose.

## 6.6.2 Early Characterizations of Public Views on Synthetic Biology in the United States

The limited data and early academic literature on social perceptions of synthetic biology in the USA offer a convergent profile of current perceptions that can be economically presented saying that (1) there is a low level of awareness, although it seems to be rapidly on the rise, (2) no visible signs of conflict have emerged at present, (3) the ethical component implicated by synthetic biology does not carry significant weight in public perceptions today, (4) but there are pointers to latent or potential reservations, beyond the conventional realm of risks (biosecurity and biosafety) and ethics, regarding the depth of intervention in nature entailed by synthetic biology (the "naturalness" frame seems to have a potentially important role in the acceptance of the area), (5) the applications likely to receive the highest level of approval in the foreseeable future are those addressing key environmental issues, and (6) there is a preference for (a) tight regulation, (b) a governance architecture based more on technical aspects than on ethical considerations and, accordingly, (c) for arrangements that give a significant say to the experts (the scientific community) versus public participation mechanisms.

A useful starting point to review the findings just mentioned is the brief report published under the title *Awareness of and Attitudes toward Nanotechnology and Synthetic Biology*, which covers the results of two pioneering studies in the United States, a national representative survey and two focus groups (Hart Research Associates 2008). Pauwels (2009) has provided a more conceptual summary of Hart's quantitative survey combined with the analysis of an additional online survey (conducted within the framework of the Cultural Cognition Project, CPP, at Yale Law School) and, particularly, an expansion of the qualitative findings contained in the Hart report. More recently, Pauwels has offered a new review of both the quantitative data and qualitative information available for the United States case (2013). We will refer here to both the Hart report and Pauwels' two review papers.

According to Pauwels, two main findings emerge from early survey research on the public's views about synthetic biology in the United States. The first is that in 2008 the adult population knew nothing at all (67 %) or little (22 %) about synthetic biology, while only 9 % knew "something" and a mere 2 % "a lot" (Hart Research Associates 2008, p. 8). Three years later, "most Americans remain unfamiliar with synthetic biology. However, despite being relatively low, public awareness of synthetic biology in the United States has nearly tripled over the past 3 years, with 26 % of the survey respondents in 2010 saying that they were aware of the topic. This is up from 22 % in 2009 and nearly three times the percentage (9 %) of those who said that they had heard about synthetic biology in 2008. Just 43 % of the respondents said that they had heard nothing at all about it, down from 67 % two years ago" (Pauwels 2013, p. 82). Secondly, notwithstanding this lack of knowledge, most individuals were willing and able to express their views on what they thought synthetic biology was about and the tradeoff between potential benefits and potential risks. In answer to the open-ended question "Regardless of how much you have heard about synthetic biology, what do you think synthetic biology is? What ideas, images, words, or phrases do you associate with synthetic biology?", the distribution of responses in 2008 was as follows: 29 % of adults admitted they didn't know; almost 29 % answered that synthetic biology was something "man-made", "artificial", "fake", "unnatural", or "not real"; 11 % said that it was about cloning, genetic engineering, or genetic manipulation; 7 % said it had to do with altering biology or biological make-up; 4 % explained it as creating artificial life; 7 % replied that it was used in medical research and another 6 % that it was used to create human skin, organs, and tissue (Hart Research Associates 2008, p. 9).[30]

These quantitative results were replicated in various focus groups, providing useful insights into the frames utilized to apprehend synthetic biology in the absence of specific knowledge about the field. The images and words associated with the development of synthetic biology were borrowed from medical or industrial activities—products such as medicines, vaccines, and plastics. The adjectives most commonly cited were consistent with the results of the open-ended question in the survey mentioned above: "man-made" was cited repeatedly, and other recurrent adjectives were "artificial", "created", "unnatural", and "synthetic". When elaborating on the goals of the technology, participants made use of the verbs "altering" and "duplicating", and the expression "improving the quality of life" was also mentioned several times. Participants in the focus groups also tended to describe synthetic biology by drawing parallels with more familiar biotechnology or biomedical developments such as cloning, genetic engineering and stem cell research (Pauwels 2009, p. 40).

Another relevant element in the framing process of emerging technologies is that applications matter (see Pardo et al. 2009). The type of synthetic biology application influences the potential reservations and specific evaluations. Among a series of potential applications, both medical and energy prospects appeared promising to the focus group participants. The top-ranked application of synthetic biology was creating new, cheaper, and cleaner sources of energy. Participants were more ambivalent about the benefits of medical applications being developed using synthetic biology, with their expectations or perceived promise (for example, eradication of genetic diseases or curing cancer) being as high as their reservations (for example, about the use of "engineered" or "re-designed" organisms in the human body) (see Pauwels 2009, p. 43).

---

[30]According to the latest survey results reviewed by Pauwels (2013), "despite their limited awareness of synthetic biology, 7 in 10 respondents reported some sense or idea about what they think synthetic biology involves, and their top of mind perceptions were focused mainly on the concept that it is human-made or artificial (30 %). Fully 12 % said that it has something to do with genetic engineering or with modifying or altering plants, crops, and cells. Smaller percentages of the respondents mentioned science or biology (6 %); cloning (6 %); machines, drugs, or advancements in medical research (5 %); or synthetic materials and chemicals (5 %). Nearly a third (29 %) of the respondents had no sense of synthetic biology or did not offer a response." (Pauwels 2013, p. 82).

Around two thirds of adult Americans offered a global evaluation of synthetic biology without proper knowledge, relying instead on associations, parallels with biotechnology and shortcuts, with 29 % believing that "benefits and risks will be about equal", 21 % thinking that "benefits will outweigh risks", 16 % that "risks will outweigh benefits", and the remaining 34 % "not sure" (Hart Research Associates 2008, p. 10). In the qualitative study, the perceived risks of synthetic biology were associated with three components: firstly, the difficulty of managing unknowns, secondly, human and, particularly, environmental side effects, and, thirdly, long-term effects. Fear of "unknowns" and "long-term effects" implies that, in the public's mind, the scientific community is not able to predict, much less control, many of the consequences of their work (Pauwels 2009, p. 44). Two further critical dimensions of public perceptions are, firstly, the issue of "who is in charge?", on the assumption that the level of security of technological develop- ment may be a function of "whose hands it is in and what they intend to do with it", and, secondly, the major evaluative axis of "not messing with God's creation", natural selection or what it is to be human; a concern deeply rooted not only in religion but also in worldviews and cultural beliefs.[31]

Pauwels' review of the empirical evidence of the CCP survey in the USA found weak support for the canonical "familiarity argument" of the public understanding of science field, that is, the tenet that support for emerging technologies will likely increase with the level of awareness of what they are about. Reported familiarity with synthetic biology was not strongly associated with respondents' perceptions of its risks and benefits. Even more significantly, the Hart survey shows that after learning about synthetic biology and being informed of its potential risks and ben- efits, the greatest shift in public opinion was toward an increase in the perception of risks. Also, how much participants knew about synthetic biology had little rela- tion to their framing of its risks and benefits. In one qualitative study reviewed, confronting the focus group participants with balanced information about syn- thetic biology did not lead to more or less support for the technology, but to a more fine-grained level of discussion. This result is consistent with previous find- ings that have linked level of informedness to more differentiation and structure in attitudes toward the object or issue and not to the valence of such attitudes (Allum et al. 2008; Evans and Durant 1995; Pardo and Calvo 2002). Finally, according to the results of the focus groups reviewed by Pauwels, most people were in favor of regulation of synthetic biology by the federal government, because of its account- ability, but also of giving a powerful voice to the scientific community and other experts due to their knowledge and competence (Pauwels 2009, p. 45).

---

[31]For interpretations of "the playing God argument" decoupled from a religious interpretation, see Peters (2006), section, "The Problem of Scientists Playing God", pp. 382–384); van den Belt (2009); and Lentzos et al. (2012), section "'Playing God' and Challenging the Organism/ Machine Divide", pp. 139–142).

## 6.6.3 Vistas on European Perceptions of Synthetic Biology

These results obtained in the first quantitative (surveys) and qualitative (focus group) studies conducted in the United States are convergent with the latest findings based on the Eurobarometer survey (Gaskell et al. 2010) carried out in 32 countries,[32] which also offers additional angles on public views of synthetic biology. The main questions covered by this international survey are: (1) awareness of synthetic biology, (2) importance of seven different criteria for taking a position on synthetic biology (from the technical and scientific aspects to the ethical component), (3) global attitudes to this area (from unconditional approval to total rejection), and (4) preferences regarding various components of the regulatory regime for synthetic biology, including 4.1, market mechanisms versus tight regulation by the government; 4.2, relative weights of scientific evidence versus ethical aspects, and 4.3, the role of experts and the public in the regulatory process. Unfortunately, there was no specific question either about a critical feature of synthetic biology, its perceived (un)naturalness, or about the perception of scientists "playing God" in trying to create life from scratch; a side effect of the weak theoretical development of the public perceptions of science field, which not infrequently leaves aside canonical variables (which should be considered a component of the "state of the art") without any explanation when designing new studies.

### 6.6.3.1 Awareness of Synthetic Biology in Europe

Since the Eurobarometer survey started with the assumption that synthetic biology is still widely unknown, respondents were first presented with the following description or semantic frame:

"Synthetic biology is a new field of research bringing together genetics, chemistry and engineering. The aim of synthetic biology is to construct completely new organisms to make new life forms that are not found in nature. Synthetic biology differs from genetic engineering in that it involves a much more fundamental redesign of an organism so that it can carry out completely new functions" (Gaskell et al. 2010, p. 120)

The level of synthetic biology's salience in the European public opinion landscape is extremely low: 8 out of 10 Europeans had not *heard* about synthetic biology before the interview took place (Table 6.6). Of those having heard about it (17 %), 50 % declared that they had never *talked* about it and 70 % that they had never *searched* for any information (see Tables 6.7 and 6.8).

---

[32]Of the 32 countries, only the 27 belonging to the EU in 2010 have been included in the analysis that follows.

**Table 6.6** Before today, have you ever heard about synthetic biology?

|       | Frequency | Percent |
|-------|-----------|---------|
| Yes   | 2219      | 16.9    |
| No    | 10924     | 83.1    |
| Total | 13143     | 100.0   |

*Source* Eurobarometer 73.1 (European Commission 2010)

**Table 6.7** Have you ever talked about synthetic biology with anyone before today?

|                        | Frequency | Percent |
|------------------------|-----------|---------|
| Yes, frequently        | 84        | 3.8     |
| Yes, occasionally      | 495       | 22.3    |
| Yes, only once or twice| 502       | 22.6    |
| No, never              | 1110      | 50.0    |
| Don't know             | 28        | 1.3     |
| Total                  | 2219      | 100.0   |

*Source* Eurobarometer 73.1 (European Commission 2010)

**Table 6.8** Have you ever searched for information about synthetic biology?

|                        | Frequency | Valid Percent |
|------------------------|-----------|---------------|
| Yes, frequently        | 70        | 3.1           |
| Yes, occasionally      | 280       | 12.6          |
| Yes, only once or twice| 303       | 13.7          |
| No, never              | 1560      | 70.3          |
| Don't know             | 6         | 0.3           |
| Total                  | 2219      | 100.0         |

*Source* Eurobarometer 73.1 (European Commission 2010)

In Europe, public awareness about synthetic biology is accordingly minimal at the beginning of the second decade of the century, far below that of biotechnology and cloning; a fact that casts doubts on the common assumption in the literature on social aspects of synthetic biology that the frame of reference should be similar to the one applied apropos of biotechnology. A legitimate question is if this very modest level of salience rules out the possibility of any study on public perceptions of what would seem to be too new an area. Not infrequently, when results from public opinion studies on objects with such a low level of visibility are reported, they are viewed as merely an artifact or as findings about a "phantom public".[33] Certainly, empirical public opinion studies of emerging and still fuzzy objects for the public's mindscape, as synthetic biology is today, must be taken as being merely exploratory, suggestive of hypotheses but unable to substantiate them (hypotheses, furthermore, based on more familiar and salient objects bearing significant similarities with the new ones under study).

---

[33]Political scientist George F. Bishop has referred to the "phantom public" when pollsters insist on measuring attitudes to objects far removed from people's attention and understanding, issues that at a particular time are not yet part of public opinion. See the chapter "Illusory Opinions on Public Affairs" (Bishop 2005, pp. 19–46).

## 6.6.3.2 Relative Weight of Different Criteria in the Evaluation of Synthetic Biology

After this caveat, the next big question is how people deal with the evaluation of synthetic biology—an object still unfamiliar to most individuals—when they are asked by interviewers to express a judgment. We know that in order to apprehend an attitudinal object, people regularly turn to shortcuts and heuristics, especially in the absence of specific knowledge and also when the object is perceived as being fuzzy in its boundaries and denotation. Using this template to determine the evaluative handles a typical European would choose in order to take a position on synthetic biology, the following hypothetical question was posed in the context of the Eurobarometer survey:

"Suppose there was a referendum about synthetic biology and you had to make up your mind whether to vote for or against. Among the following, what would be the most important issue about which you would like to know more?" The question offered the following seven categories (each of them could be selected as a 1st, 2nd, and 3rd response): (1) scientific and technical aspects, (2) source of funding of the research, (3) potential risks, (4) potential benefits, (5) distribution of risks and benefits, (6) regulation and control, and (7) social and ethical aspects. The response distribution of the first mention (which captures the most important or salient aspect as perceived by the public) shows that the potential risks and benefits related to synthetic biology are of prime importance to respondents (see Table 6.9). This result suggests that, as in most other cases of new technoscientific developments, people tend to use the evaluative axis of the risks-benefits tradeoff. Taking into account the three categories chosen by most people, all the other informational aspects presented were of interest to a significant proportion of the European public, although risks, benefits and the distribution of both occupied the top positions. A third, significant conclusion is that the actual social and ethical issues raised by synthetic biology do not appear, currently, to be particularly relevant for most people in the majority of countries (reaching only 16 % of mentions after adding the three responses, although in a few societies the concerns expressed are of a considerable magnitude). The basic template of "moral-immoral" seems, for most people, to be distant from the synthetic biology object, possibly because there is no clear involvement of humans and/or animals as primary subjects in this type of research, that is, due to the absence of the main (relationships between humans) and even secondary (behavior of humans in relation to animals) targets of moral concern and ethical analysis in Western cultures.

There are significant national differences in the European Union regarding the role assigned to social and ethical aspects in the evaluation of synthetic biology. According to first mentions of "social and ethical issues" from the nine possibilities, we can see that in three countries there are significant minorities who care about them very much: Denmark (11 %), the Netherlands (10 %) and Sweden (8 %). And, combining the three possible mentions, we find five societies where a considerable subset of the population rates these issues as very important: namely, the Netherlands (39 %), Denmark (37 %), Sweden (33 %), Belgium (25 %), and

**Table 6.9** Aspects of synthetic biology about which respondents would like to know more

| | Frequency 1st mention | Percent 1st mention | Frequency 1st, 2nd, 3rd mention | Percent 1st, 2nd, 3rd mention |
|---|---|---|---|---|
| What the scientific processes and techniques are | 1982 | 15.1 | 4134 | 31.5 |
| Who is funding the research and why | 1045 | 7.9 | 3114 | 23.7 |
| What the claimed benefits are | 2799 | 21.3 | 6809 | 51.8 |
| What the possible risks are | 3113 | 23.7 | 8371 | 63.0 |
| Who will benefit and who will bear the risks | 1315 | 10.0 | 5313 | 40.4 |
| What is being done to regulate and control synthetic biology | 668 | 5.1 | 3754 | 22.6 |
| What is being done to deal with the social and ethical issues involved | 441 | 3.4 | 2113 | 16.1 |
| Other (SPONTANEOUS) | 26 | 0.2 | 94 | 0.7 |
| None (SPONTANEOUS) | 427 | 3.3 | 597 | 4.7 |
| Don't know | 1327 | 10.1 | 1563 | 11.9 |
| Total | 13143 | 100.0 | | |

*Source* Eurobarometer 73.1 (European Commission 2010)

Great Britain (22 %). In these countries at least, the scientific community working in synthetic biology should pay attention to maintaining an open mind about the potential reservations and views the public may have.

### 6.6.3.3 Global Positions on the Acceptance of Synthetic Biology

An additional "vista" on public perceptions of synthetic biology can be obtained from the responses to a hypothetical scenario of approval-disapproval. Specifically, the question posed was "Overall, what would you say about synthetic biology?" and the possible responses were: (1) full approval, no special laws necessary, (2) approval, regulated by strict laws, (3) no approval, except under very special circumstances, (4) no approval under any circumstances, and (5) don't know. As the distribution in Table 6.10 shows, at present, under conditions of a limited level of awareness and information about synthetic biology, a relative majority would be inclined to approve it, provided strict regulations are introduced. However, another large group would demand even stronger constraints

**Table 6.10** Approval of synthetic biology

|  | Frequency | Percent |
|---|---|---|
| You fully approve and do not think that special laws are necessary | 448 | 3.4 |
| You approve as long as this is regulated by strict laws | 4783 | 36.4 |
| You do not approve except under very special circumstances | 2744 | 20.9 |
| You do not approve under any circumstances | 2211 | 16.8 |
| Don't know | 2956 | 22.5 |
| Total | 13142 | 100.0 |

*Source* Eurobarometer 73.1 (European Commission 2010)

(suggested by the clause "except under very special circumstances"). A small but significant minority (17 %) would favor a ban on this area of research and applications. A large percentage (23 %) doesn't have a position yet about the hypothetical course of action to take. The main interpretation of the distribution of responses is that synthetic biology is a potentially controversial technology, with the capacity to generate division. It also seems, however, that there is no crystallized opposition at present and, contingent on an appropriate regulatory framework, a slight majority would be inclined to consent to its development. Focusing only on the two polarized positions ("fully approve, no special laws necessary" and "do not approve under any circumstances"), in all European countries the segment favoring non-approval is much larger than the one willing to accept synthetic biology without special constraints. This extremely critical segment is particularly numerous in East Germany (28 %), Greece (26 %), Austria (23 %), Denmark (21 %) and the Netherlands (21 %). In the country with the highest full acceptance of synthetic biology, Spain, only 6 % would support this position.

### 6.6.3.4 The Regulatory Regime for Synthetic Biology

The governance of synthetic biology—much as in other emerging scientific and technological areas—aims to balance measures for the promotion of various values and goals: (1) promoting scientific freedom and progress, that is, setting the scene for research, development, innovation, and resulting benefits for society (most often interpreted as tackling grand challenges, or leading to more resources and economic growth), (2) avoiding risks, that is, ensuring safety and security in the areas of human health, the natural environment, and biosecurity, and (3) respect for moral views and other central values that usually vary within and across societies.

Similar sets of goals are reflected in the current European Commission science funding policy: the concept of "Responsible Research and Innovation" is meant to promote research and innovation that drives towards human benefit and societal desirability, is sustainable by avoiding significant adverse effects through effective risk management, and is ethically acceptable (European Commission 2011, 2012).

It is extremely challenging to balance different values that may be inherently conflicting. There are opposing interests, matters of priority, and diverging opinions about the relative importance of diverse goals and about how this balancing could be best accomplished. There has thus been a lively debate about the future governance of synthetic biology. The historical background has also played an important role in the comparatively early introduction of scientific and policy initiatives, which has mainly been motivated by a perceived need to avoid the dynamics seen in biotechnology debates in the past.

Scientists' bottom-up governance initiatives have largely focused on standardization, biosecurity, biosafety and avoidance of public unease (Check 2006; Maurer et al. 2006; Tepfer 2005). While self-regulation undoubtedly has a major role to play in some areas of the governance of synthetic biology, these foci have prompted concerted reactions by civil society organizations, which have perceived them as attempts to avoid new governmental regulation and broader socioeconomic and political issues (Pollack 2010; Torgersen 2009). To some extent, the synthetic biology community has accommodated such reactions by offering opportunities at professional meetings to address wider societal issues than those of safety and security. For example, human practices and ethics are commonly addressed in The International Genetically Engineered Machine (iGEM) Foundation's competitions, and the ETC Group contributed to "SB4.0"—the 2008 Synthetic Biology Conference in the BioBricks Foundation's Series "SBx.0".

Scientists and science organizations continue to actively shape policy in synthetic biology.[34] One example is the "Six party symposia on synthetic biology", chaired by key players in the field, where the focus at the final 2012 meeting was explicitly on "next-generation tools, platforms, and infrastructure necessary for continued progress in synthetic biology and the associated policy implications" (Joyce et al. 2013, p. 6). This way of putting it already shows that the progress of scientific research remains a constant priority. There is thus still reason to doubt motivations, as illustrated also by an inspection of many ELSA activities.

Like initiatives emanating from the scientific community, policy initiatives calling for research on ELSA and public engagement have also been part of the synthetic biology field from the start, and synthetic biologists have been pro-active in supporting such activities. There have been numerous projects at technology assessment and policy analysis institutions, and the topic has received substantial support from national and international research organizations.[35] As Gottweis (2008) has pointed out, participation and governance have become intermingled to an unusual extent in the domain of the "politics of life", because a diversity of values as well as feelings of distrust and uncertainty have to be accommodated in an agenda of openness, participation and accountability. In fact, openness, accountability and participation, alongside effectiveness and coherence, were defined as the main principles of "good governance" by the European Commission in 2001

---

[34]For an overview, see the second part of Acevedo-Rocha (2016).

[35]For an overview, see Zhang et al. (2011).

(COM 2001). Previous ELSA activities had been heavily criticized for effectively reducing participatory events to public relations activities and advertisement (e.g., Bogner and Menz 2005). In synthetic biology, this criticism was taken into account at an early stage and led to a focus on upstream engagement of the public and on securing analytical input from the social sciences and humanities (Calvert and Martin 2009; Torgersen 2009).

However, reality may not be quite as rosy as appears on first sight. Two opposing streams of criticism can be identified: (1) according to some critics, current approaches over-emphasize the role of the public, give too much power to non-governmental organizations (NGOs), demote the science component, and may ultimately hinder innovation and development (Lowrie and Tait 2011; Tait 2012), and (2) despite past criticism of some ELSA research serving public relations efforts rather than critical analysis, current approaches still sometimes amount to fig-leaf activities: the synthetic biology scientific community and government institutions with a focus on innovation for economic growth have, at least in some cases, appeared to instrumentalize social scientists and public participation initiatives to promote public acceptance (cf. Marris 2014).

These opposing views may be related to different underlying values and worldviews. In both cases, the need to integrate various governance strategies is acknowledged, but the type of strategy differs as does the weight given to the different interests and values (Garfinkel et al. 2007; Weir and Selgelid 2009). Controversies revolve around the roles of scientists, industry, governmental and trans-national regulators, NGOs/community service organizations and other stakeholders and citizen groups. How should decisions about funding and changes in legal frameworks be made? How much weight should be given to the various interests, and how much weight should be given to science versus values? A key component of the answer to these questions is the views of the public, since in the end good governance mechanisms require a certain level of consent from society at large. In the following pages of this section, we examine the main profile of those views.

Consistent with the scenario of qualified approval of synthetic biology dependent on regulation based on "strict laws", a very large majority in Europe would favor the government having a strong role versus giving synthetic biology freedom to operate in the market like any conventional business (77 % versus 11 % respectively, see Table 6.11). A second aspect of the regulatory regime of a scientific area like synthetic biology is the relative weight of strictly technical considerations

**Table 6.11** Preference for regulation by the government versus market mechanism

|  | Frequency | Percent |
|---|---|---|
| Synthetic biology should be tightly regulated by the government | 10119 | 77.0 |
| Synthetic biology should be allowed to operate in the market place like a business | 1412 | 10.7 |
| Don't know | 1611 | 12.3 |
| Total | 13142 | 100.0 |

*Source* Eurobarometer 73.1 (European Commission 2010)

and, at the other extreme, of moral and ethical values. A large majority favors scientific evidence (52 %), while a third of the population gives more weight to moral criteria (34 %) and around 14 % have no opinion (see Table 6.12). Aligned with this preference, an even higher percentage of the population is inclined to rely on experts rather than following what the public thinks in order to reach a decision about synthetic biology (see Table 6.13).

Gaskell and the co-authors of the report *Europeans and Biotechnology in 2010. Winds of change?* (p. 9) have proposed an intriguing typology of governance, based on the combined responses to the nature of the criteria to be employed (for example, scientific or moral) and the main input that should inform decisions about synthetic biology (for example, the views of experts or of the public). The first type is called the "principle of scientific delegation" and relies on scientific criteria and expert input. According to Gaskell and co-authors, an institutional correlate would be expert commissions on risk assessment, although many other arrangements could be compatible with the preference for technical competence and decision-making informed by the experts. The second is called the "principle of scientific deliberation" and combines scientific evidence and public "voice". Consensus conferences would be one of the corresponding institutional forms. The third is the "principle of moral delegation" (ethics, experts) and its organizational embodiment could be "ethics committees". The last is the "principle of moral deliberation", and relies on the moral dimension and the views of the public, which would correspond with various forms of public deliberation such as people's initiatives (see Table 6.14).

**Table 6.12**  Approval of synthetic biology: scientific evidence versus moral criteria

|  | Frequency | Percent |
|---|---|---|
| Decisions about synthetic biology should be based primarily on scientific evidence | 6888 | 52.4 |
| Decisions about synthetic biology should be based primarily on the moral and ethical issues | 4427 | 33.7 |
| Don't know | 1827 | 13.9 |
| Total | 13142 | 100.0 |

*Source* Eurobarometer 73.1 (European Commission 2010)

**Table 6.13**  Approval of synthetic biology: experts versus lay public

|  | Frequency | Percent |
|---|---|---|
| Decisions about synthetic biology should be based mainly on the advice of experts | 7725 | 58.8 |
| Decisions about synthetic biology should be based mainly on what the majority of people in a country thinks | 3817 | 29.0 |
| Don't know | 1600 | 12.2 |
| Total | 13142 | 100.0 |

*Source* Eurobarometer 73.1 (European Commission 2010)

**Table 6.14** Principles of governance of synthetic biology

|  | Frequency | Percent |
|---|---|---|
| Scientific delegation | 5289 | 51.6 |
| Moral delegation | 1572 | 15.3 |
| Scientific deliberation | 1027 | 10.0 |
| Moral deliberation | 2357 | 23.0 |
| Total | 10245 | 100.0 |

*Source* Eurobarometer 73.1 (European Commission 2010)

The two most interpretable of the four principles proposed by Gaskell and colleagues are the first and the last, taking into account the questions asked and the fact that most people are totally unaware of the main institutional forms of the other two. A clear majority of the adult population in Europe would be happy with decisions about synthetic biology taken on the basis of scientific reasons, and relying on bodies that follow technical criteria versus the views of the public at large ("scientific delegation", according to Gaskell's label). Fifty-two per cent of the population in the 27 EU countries in 2010 would prefer this option, which implies a high level of confidence in professional competence and scientific knowledge. The socio-demographic profile or segmentation of people favoring each of these two options is as follows. The principle of "scientific delegation" is more prevalent among men, the 35 to 44 year age group, and those with more formal education, the non-religious and those identifying with the left on an ideological scale. The principle of "moral deliberation" is the option best represented among women, the 65 year and over age group, those with less formal education, and religious individuals (both Catholics and Protestants). Focusing on educational level alone, we can see a clear pattern emerging. Individuals with more years of formal education are the ones that most favor "scientific delegation": 44 % of people that left full-time education when they were 15 years old, 52 % of those that left education when they were 16 to 19 years old, and 56 % of people with a university degree (aged 20 or more when they finished education). At the opposite extreme, a little more than a fifth of the population (23 %) favor the use of ethical criteria and the views of society versus scientific considerations and expert views. People with less formal education tend to prefer the option of public involvement and give more weight to ethical considerations: 33 % of those aged 15 years or under, 23 % in the 16 to 19 year age group and 17 % of those aged 20 years and over.

As in the case of most questions that are culturally shaped, there is large variability in these preferences among the European societies studied, with countries such as Denmark, Austria and the Netherlands in which just a third of the population prefers a scheme of governance based on "scientific delegation", and countries such as Hungary, Spain, the Czech Republic and Italy where at least 60 % of the population embrace this way of regulating synthetic biology. The states belonging to the former West and East Germany have a diverse profile: in both cases there is sharp division, but in the East there is a larger segment favoring "scientific delegation" over "moral delegation" (Table 6.15).

**Table 6.15** Principles of governance of synthetic biology by country

| Countries | Scientific Delegation | Moral Delegation |
|---|---|---|
| Hungary | 66 | 13 |
| Spain | 62 | 18 |
| Czech Republic | 62 | 19 |
| Italy | 61 | 22 |
| Belgium | 57 | 14 |
| France | 55 | 16 |
| Great Britain | 55 | 20 |
| Portugal | 52 | 15 |
| Sweden | 52 | 15 |
| Poland | 50 | 23 |
| Greece | 46 | 38 |
| Germany (East) | 38 | 33 |
| Netherlands | 38 | 32 |
| Austria | 38 | 32 |
| Denmark | 37 | 26 |
| Germany (West) | 30 | 39 |

*Source* Gaskell et al. 2010

## 6.6.4 Synthetic Biology, Unnaturalness and "Playing God"

A potentially very important angle for the evaluation of synthetic biology is the perceived naturalness (or unnaturalness) of its different applications, which is a function of the depth and scope of intervention by science in the natural domain. "Naturalness" today is one of the main elements of a worldview or cultural vector associated with the evaluation of a large series of processes and objects.[36] The dominant worldview about nature encompasses attributes such as beauty, peace, fragility and interconnectedness, and also skepticism or strong reservations about the modification of natural processes by science and technology, which is perceived as being equivalent to tampering with the "wisdom" embedded in millions of years of evolution, of piecemeal trial and error. Most people worry about the level of degradation of nature, as reported in a constant flow of published research in scientific areas like ecology and conservation biology and corroborated by personal macroscopic experience in everyday life. In a number of domains, mainly agriculture and food production, the attribute of naturalness carries a positive valence both for the environment and human health (Rozin et al. 2012). Even if people are not consistent in their everyday behavior with their beliefs about nature, it remains a significant fact that most individuals, since the closing decades of the 20th century, would like to preserve nature as much as possible, even while many of them are aware that most facets of "nature", including today's landscapes,

---

[36]For a systematic and enlightening analysis of the naturalness concept, see Siipi (2005).

reflect a process of systematic modification by humans in the past. Unless the potential gains are fundamental, a large majority would prefer not to pursue the path of aggressive intervention in nature, particularly if this entails the modification of the genetic blueprint. This is the main cultural reason for resistance to or ambivalence about biotechnology.

Not only one of the main goals but also the explicit narrative of synthetic biology involves radically surpassing and ameliorating nature, or even entirely bypassing known biological processes in order to satisfy needs in areas such as energy production, bioremediation and health. In no other area of science can we find a more fundamental and radical role for the figure of the scientist than in synthetic biology. It is not surprising that the classical symbolic figures of Faust and Frankenstein and the perception of scientists as "playing God" have resurfaced in the context of synthetic biology. The idea of the researcher "playing God" is, at the beginning of this century, largely disconnected from a religious or theological interpretation, as suggested by the large number of non-believers and largely secular societies such as France that use this perspective in assigning a negative valence to radical modifications of nature.[37]

Synthetic biology has embedded and promotes an ontology populated by new objects and entities that seem to radically cross the boundaries between life and not-life, animate and inanimate, the natural and the artificial (Lentzos et al. 2012). The first systematic and robust study seeking to ascertain the role of the "unnatural objection" in the perception of synthetic biology, the work of Dragjlovic and Einsiedel, has convincingly shown that concerns over the use of genetic material of dissimilar organisms ("evolutionary distance" between "donor" and "created organisms"), views of nature and, particularly, the unnaturalness frame, if activated, would play a significant role in the evaluation of synthetic biology (Dragojlovic and Einsiedel 2013). Although the public at large is currently unaware of the development and goals of synthetic biology, it is plausible that, if the organizations opposing it were able to activate and give salience to the frame of extreme "unnaturalness", most individuals would oppose many of the applications except for those few that address critical social challenges in areas like the natural environment and animal and human health. The scientific community should be aware of this overarching cultural dimension of their work, which is beyond the bounds of conventional ethical and risk analyses.

---

[37]Dabrock in his insightful paper "Playing God? Synthetic biology as a theological and ethical challenge" (2009), in our view misses the point that "playing God" could and does have a non-religious interpretation, even if this cultural angle has some echoes of its religious origin. In contrast, van den Belt, in his "Playing God in Frankenstein's Footsteps: Synthetic Biology and the Meaning of Life" (2009) offers a cogent secular meaning of that notion.

## 6.7  Conclusion

Public voices play a significant role today in shaping the regulatory climate and trajectory of some subsets of science and technology, particularly in those cases that have the potential to disrupt or collide with social values and perceptions. It is no longer valid to dismiss public concerns, fears and preferences regarding specific scientific developments on the basis of a low level of scientific literacy, real or alleged. Certainly, the diffusion of scientific knowledge to the public at large is in need of renewed attention as regards its agents, scope, methods, narratives and channels, and the scientific community should, of course, incorporate that task in its portfolio of recurrent activities. But much more is needed. Above all, a more balanced interaction between experts and the public, alongside a conscious engagement of the scientific community with the social values of the host society.

The experience with biotechnology in many advanced societies has shown that cultural factors could indirectly affect the regulatory treatment of biotechnological applications in sensitive areas such as food and agriculture, and, even more importantly, the public's willingness to adopt them. Besides the tradeoff between goals and means, a number of vectors influence the valence (positive, negative or ambivalent) of public attitudes and behaviors, among them risk perceptions, ethical views and trust in the scientific community and the institutional architecture of regulation. Beyond that, other powerful cultural variables (worldviews, frames, values and symbols) associated with attitudes to subsets of the life sciences should be considered. One of the central worldviews (with an array of connected values) that affects such areas of science is the current vision of nature and the vector "natural"–"unnatural". At the beginning of the 21st century, a large majority of the population is aware of the negative environmental externalities of a model of growth that has made intensive use of science and technology while, in turn, permeating the goals and subculture of a large subset of the scientific community. Accordingly, many individuals are acutely sensitive to new scientific knowledge that pushes forward the boundaries of "natural" processes and objects, particularly if such knowledge involves the alteration of the genetic blueprint of life.

Synthetic biology has a practical and also a cultural component in potential conflict with the dominant worldview about nature, and the public's preference for "naturalness" in many areas. The explicit and embedded general narrative about life, its radical modification and even creation (the so-called "regenesis") developed by leading synthetic biology researchers could trigger strong resistance on the part of the public and compromise the future trajectory of the field. And, as pointed out by Gerd Winter (see legal chapter in this book), "the cultural concerns" of the public should "be accepted as a legitimate ground for regulatory restrictions". At present, the worldview about the natural domain and the unnaturalness frame are dormant in relation to synthetic biology, due to the low salience attained to date by this emergent scientific area. However, given the appropriate conditions (visibility and activation of the worldview and frame by environmental organizations and other groups), they may prove to be a formidable barrier to research and, particularly, to the deployment of applications that do not address critically important goals. As well as the factor of the synthetic biology-enabled evolutionary distance between the donor and recipient of traits and parts, the social reception of the corresponding applications would depend very much on whether these were perceived as creating objects and entities comparable to physical artifacts (where there would be less or no resistance), or as biological parts and systems (where there would be greater resistance). Beyond conventional ELSI analyses, the scientific community working in this area should build a minimalistic narrative of its goals, and develop a research culture that takes into account the current mindset regarding nature and "(un)naturalness" and the role of science in its modification, creation and management.

# References

Acevedo-Rocha CG (2016) The synthetic nature of biology. In: Hagen K, Engelhard M, Toepfer G (eds) Ambivalences of creating life. Societal and philosophical dimensions of synthetic biology. Springer, Berlin, pp 9–53

Allum N, Sturgis P, Tabourazi D, Brunton-Smitz I (2008) Science knowledge and attitudes across cultures: a meta-analysis. Public Underst Sci 17:35–54

Bauer MW, Allum N, Miller S (2007) What can we learn from 25 years of PUS survey research? Liberating and expanding the agenda. Public Underst Sci 16:79–95

Bauer MW, Gaskell G (2002) Biotechnology. The making of a global controversy. CUP, Cambridge

Bauer MW, Shukla R, Allum N (eds) (2012) The culture of science. How the public relates to science across the globe. Routledge, New York

Ben-David J (1984) The scientist's role in society. A comparative study. Chicago University Press, Chicago

Birnbacher D (1999) Ethics and social science: what kind of cooperation? Ethical Theory Moral Pract 2:319–336

Bishop GF (2005) The illusion of public opinion. Fact and artifact in American public opinion polls. Rowman and Littlefield Publishers, Inc., New York

Bodmer W (2010) The public understanding of science, the BA, the royal society and COPUS. Notes Rec R Soc. doi:10.1098/rsnr.2010.0035

Bogner A, Menz W (2005) Alternative Rationalitäten? Technikbewertung durch Laien und Experten am Beispiel der Biomedizin. In: Bora A, Decker M, Grunwald A, Renn O (eds) Technik in einer fragilen Welt. Die Rolle der Technikfolgenabschätzung, edition sigma, Berlin, pp 383–391

Boldt J, Müller O (2008) Newtons of the leaves of grass. Nat Biotechnol 26:387–389. doi:10.103 8/nbt0408-387

Bowler PJ, Morus IR (2005) Making modern science. A historical survey. The University of Chicago Press, Chicago

Calvert J, Martin P (2009) The role of social scientists in synthetic biology 10:201–204

Check E (2006) Synthetic biologists try to calm fears. Nature 441:388–389

Church G, Regis E (2012) Regenesis. How synthetic biology will reinvent nature and ourselves. Basic Books, New York

Cobb MD, Macoubrie J (2004) Public perceptions about nanotechnology: Risks, benefits and trust. J Nanoparticle Res 6:395–405

COM (Commission of the European Communities) (2001) European governance. A White Paper. COM (2001) 428 final, Brussels

Daamen DDL, Van der Lans IA, Midden CJH (1990) Cognitive structures in the perception of modern technologies. Sci Technol Hum Values 15:202–225. doi:10.1177/016224399001500203

Dabrock P (2009) Playing God? Synthetic biology as a theological and ethical challenge. Syst Synth Biol 3:47–54. doi:10.1007/s11693-009-9028-5

Dragojlovic N, Einsiedel E (2013) Framing synthetic biology. Sci Commun 35:547–571. doi:10.1177/1075547012470707

Einsiedel E (2005) In the Public Eye: The Early Landscape of Nanotechnology among Canadian and US Publics. AZoNano Online J Nanotechnol 1:1–10

Einsiedel E, Kamara MW, Boy D, et al. (2006) The coming of age of public participation. In: Gaskell G, Bauer MW (eds) Genomics and society: legal, ethical and social dimensions. Earthscan, London, pp 95–112

ETC Group (2007) Extreme genetic engineering: an introduction to synthetic biology. Ottawa

European Commission (2010) Eurobarometer 73.1 on the Life Sciences and Biotechnology. http://ec.europa.eu/public_opinion/archives/ebs/ebs_341_en.pdf. Accessed 24 July 2015

European Commission (2011) towards responsible research and innovation in the information and communication technologies and security technologies fields. Publications Office of the European Union, Luxembourg. doi:10.2777/58723

European Commission (2012) Ethical and regulatory challenges to science and research policy at the global level. Publications Office of the European Union, Luxembourg. doi:10.2777/35203

Evans G, Durant J (1995) The relationship between knowledge and attitudes in the public understanding of science in Britain. Public Underst Sci 4:57–74. doi:10.1088/0963-6625/4/1/004

Ganguli-Mitra A, Schmidt M, Torgersen H, Deplazes A, Biller-Andorno N (2009) Of Newtons and heretics. Nat Biotechnol 27:321–322. doi:10.1038/nbt0409-321

Garfinkel MS, Endy D, Epstein GL, Friedman RM (2007) Synthetic genomics: options for governance. The J Craig Venter Institute, Rockville

Gaskell G, Allansdottir A, Allum N, Fischler C, Hampel J, Jackson J, Kronberger N, Mejlgaard N, Revuelta G, Schreiner C, Stares S, Torgersen H, Wagner W (2006) Europeans and biotechnology in 2005: patterns and trends. A report to the European Commission's Directorate General for Research. European Commission, Brussels

Gaskell G, Stares S, Allansdottir A, Allum N, Castro P, Esmer Y, Fischler C, Jackson J, Kronberger N, Hampel J, Mejlgaard N, Quintanilha A, Rammer A, Revuelta G, Stoneman P, Torgersen H, Wagner W (2010) Europeans and biotechnology in 2010 winds of change? A report to the European Commission's Directorate-General for Research. European Commission, Brussels

Gigerenzer G (2007) Gut feelings. The intelligence of the unconscious. Penguin Books, London

Gottweis H (2008) Participation and the new governance of life. Biosocieties 3:265–286. doi:10.1017/S1745855208006194

Gschmeidler B, Seiringer A (2012) Knight in shining armour" or "Frankenstein's creation"? The coverage of synthetic biology in German-language media. Public Underst Sci 21:163–173. doi:10.1177/0963662511403876

Hård M, Jamison A (1988) The Intellectual Appropriation of Technology. The MIT Press, Cambridge

Hart Research Associates (2008) Awareness of and attitudes toward nanotechnology and synthetic biology: a report of findings based on a national survey among adults. Washington

Heise UK (2004) Science, Technology, and Postmodernism. In: Connor S (ed) Cambridge companion to postmodernism. CUP, Cambridge, pp 136–167

Hirschman AO (1970) Exit, voice, and loyalty. Harvard University Press, Cambridge

Holton G (1995) Einstein, history, and other passions. American Institute of Physics, Woodbury

Holton G (1988) Thematic origins of scientific thought. Kepler to Einstein. Harvard University Press, Cambridge

Holton G (1978) The scientific imagination. Case studies. Cambridge University Press, Cambridge

House of Lords, Select Committee on Science and Technology (2000) Science and society. The Stationery Office, London

Hugues TP (2004) American genesis, 2nd edn. The University of Chicago Press, Chicago

Joyce S, Mazza A-M, Kendall S (Rapporteurs), Committee on science, technology, and law; policy and global affairs; Board on life sciences; division on earth and life sciences; National Academy of Engineering; National Research Council (2013) Positioning synthetic biology to meet the challenges of the 21st century: summary report of a six academies symposium series. National Academies Press

Kaiser M (2012) Commentary: looking for conflict and finding none? Public Underst Sci 21:188–194

Keller EF (1995) Refiguring life. Metaphors of twentieth-century biology. Columbia University Press, New York

Kronberger N, Holtz P, Kerbe W, Strasser E, Wagner W (2009) Communicating synthetic biology: from the lab via the media to the broader public. Syst Synth Biol 3:19–26. doi:10.1007/s11693-009-9031-x

Lassen J, Gjerris M, Sandøe P (2006) After Dolly—ethical limits to the use of biotechnology on farm animals. Theriogenology 65:992–1004

Lentzos F, Cockerton C, Finlay S, Hamilton A, Zhang J, Rose N (2012) The societal impact of synthetic biology. In: Freemont PS, Kitney RI (eds) Synthetic biology: a primer. World Scientific, Singapore, pp 131–149

Levitt M, Weiner K, Goodacre J (2005) Gene Week: a novel way of consulting the public. Public Underst Sci 14:67–79

Lévy-Leblond J (1992) About misunderstandings about misunderstandings. Public Underst Sci 1:17–21. doi:10.1088/0963-6625/1/1/004

Lowrie H, Tait J (2011) Guidelines for appropriate risk governance of synthetic biology. Int Risk Gov Counc Policy Br. http://www.irgc.org/IMG/pdf/irgc_SB_final_07jan_web.pdf. Accessed 10 Feb 2014

Marris C (2014) The construction of imaginaries of the public as a threat to synthetic biology. Sci Cult 24:83–98

Maurer SM, Lucas KV, Terrell S (2006) From understanding to action. Community-based options for improving safety and security in synthetic biology. University of California, Berkeley

Marx L (1988) The pilot and the passenger: essays on literature, technology, and culture in the United States. Oxford University Press, Oxford

Marx L (2001) The domination of nature and the redefinition of progress. In: Marx L, Mazlish B (eds) Progress. Fact or illusion. The University of Michigan Press, Ann Arbor, pp 201–218

Mazur A (1981) The dynamics of technical controversy. Communications Press Inc, Washington

Miller JD, Pardo R (2000) Civic scientific literacy and attitude to science and technology: a comparative analysis of the European Union, the United States, Japan, and Canada. In: Dierkes M, von Grote C (eds) Between understanding and trust: the public, science and technology. Harwood Academic Publishers, Amsterdam, pp 131–156

Miller JD, Pardo R, Niwa F (1997) Public perceptions of science and technology. A comparative study of the European Union, the United States, Japan, and Canada. Fundación BBV-Chicago Academy of Sciences, Bilbao

Molewijk B, Stiggelbout AM, Otten W, Dupuis HM, Kievit J (2004) Empirical data and moral theory. A plea for integrated empirical ethics. Med Heal Care Philos 7:55–69

Nielsen AP, Lassen J, Sandøe P (2011) Public participation: democratic ideal or pragmatic tool? The cases of GM foods and functional foods. Public Underst Sci 20:163–178

Norris P (1999) Critical citizens. Global support for democratic government. Oxford University Press, Oxford

Nye JSJ, Zelikow PD, King DC (1997) Why people don't trust government. Harvard University Press, Cambridge

Pardo R (2012) Worldviews, frames, trust and perceptions of stem cells across Europe. In: Bauer MW, Shukla R, Allum N (eds) The culture of science. How the public relates to science across the globe. Routledge, New York, pp 353–372

Pardo R, Calvo F (2002) Attitudes toward science among the European public: a methodological analysis. Public Underst Sci 11:155–195. doi:10.1177/0963662202129084859

Pardo R, Calvo F (2006a) Mapping perceptions of science in end-of-century Europe. Sci Commun 28:3–46

Pardo R, Calvo F (2006b) Are Europeans really antagonistic to biotech? Nat Biotechnol 24:393–395

Pardo R, Calvo F (2008) Attitudes toward embryo research, worldviews, and the moral status of the embryo frame. Sci Commun 30:8–47

Pardo R, Engelhard M, Hagen K, Jørgensen RB, Rehbinder E, Schnieke A, Szmulewicz M, Thiele F (2009) The role of means and goals in technology acceptance. A differentiated landscape of public perceptions of pharming. EMBO Rep 10:1069–1075. doi:10.1038/embor.2009.208

Pardo R, Midden C, Miller JD (2002) Attitudes toward biotechnology in the European Union. J Biotechnol 98:9–24

Pauwels E (2009) Review of quantitative and qualitative studies on US public perceptions of synthetic biology. Syst Synth Biol 3:37–46. doi:10.1007/s11693-009-9035-6

Pauwels E (2013) Public understanding of synthetic biology. Bioscience 63:79–89. doi:10.1525/bio.2013.63.2.4

Pei L, Gaisser S, Schmidt M (2011) Synthetic biology in the view of European public funding organisations. Public Underst Sci 21:149–162. doi:10.1177/0963662510393624

Pepper D (1996) Modern environmentalism. Routledge, London

Peters T (2006) Contributions from practical theology and ethics. In: Clayton P (ed) Oxford Handbook of Religion and Science. Oxford University Press, Oxford, pp 372–387

Pollack A (2010) Synthetic biology does not need regulation now, panel says. The New York times. Accessed 16 Dec 2010

Priest SH, Bonfadelli H, Rusanen M (2003) The "trust gap" hypothesis: predicting support for biotechnology across national cultures as a function of trust in actors. Risk Anal 23:751–766

Rehbinder E, Engelhard M, Hagen K, Jørgensen RB, Pardo-Avellaneda R, Schnieke A, Thiele F (2009) Pharming. Promises and risks of biopharmaceuticals derived from genetically modified plants and animals. Springer, Berlin

Rowe G, Frewer LJ (2000) Public participation methods: a framework for evaluation. Sci Technol Hum Values 25:3–29

Rowe G, Frewer LJ (2004) Evaluating public participation exercises: a research agenda. Sci Technol Hum Values 29:512–556

Rowe G, Frewer LJ (2005) A typology of public engagement mechanisms. Sci Technol Hum Values 30:251–290

Rowe G, Marsh R, Frewer LJ (2004) Evaluation of a deliberative conference. Sci Technol Human Values 29:88–121

Rozin P, Fischler C, Shields-Argelès C (2012) European and American perspectives on the meaning of natural. Appetite 59:448–455. doi:10.1016/j.appet.2012.06.001

Siegrist M (2000) The influence of trust and perceptions of risks and benefits on the acceptance of gene technology. Risk Anal 20:195–204

Siipi H (2005) Naturalness, unnaturalness, and artifactuality in bioethical argumentation. University of Turku (Reports from the Department of Philosophy), Turku

Sjöberg L (2004) Principles of risk perception applied to gene technology. EMBO Rep 5:S47–S51. doi:10.1038/sj.embor.7400258

Slovic P (2000) The perception of risk. Earthscan, London

Sugarman J (2004) The future of empirical research in bioethics. J Law Med Ethics 32:226–231

Tait J (2012) Adaptive governance of synthetic biology. EMBO Rep 13:579. doi:10.1038/embor.2012.76

Taylor SE (1981) The interface of cognitive and social psychology. In: Harvey J (ed) Cognition, social behavior, and the environment. Erlbaum, Hillsdale, pp 88–114

Tepfer M (2005) How synthetic biology can avoid GMO-style conflicts. Nature 437:476

The Royal Society of London (1985) The public understanding of science. The Royal Society, London

Torgersen H (2009) Synthetic biology in society: learning from past experience? Syst Synth Biol 3:9–17. doi:10.1007/s11693-009-9030-y

Torgersen H, Hampel J (2012) Calling controversy: assessing synthetic biology's conflict potential. Public Underst Sci 21:134–148. doi:10.1177/0963662510389266

Van den Belt H (2009) Playing god in frankenstein's footsteps: synthetic biology and the meaning of life. Nanoethics 3:257–268. doi:10.1007/s11569-009-0079-6

Weir L, Selgelid MJ (2009) Professionalization as a governance strategy for synthetic biology. Syst Synth Biol 3:91–97. doi:10.1007/s11693-009-9037-4

White L Jr (1967) The historical roots of our ecological crisis. Science 155(3767):1203–1207. doi:10.1126/science.155.3767.1203

Worsley P (1987) The new introducing sociology. Penguin Books, Harmondsworth

Zhang JY, Marris C, Rose N (2011) The transnational governance of synthetic biology: scientific uncertainty, cross-borderness and the 'art' of governance. BIOS Working Paper, BIOS, London School of Economics and Political Science, London

Zerubavel E (1997) Social mindscapes. An invitation to cognitive sociology. Harvard University Press, Cambridge

# Chapter 7
# In Search for a Legal Framework for Synthetic Biology

Gerd Winter

> How fleeting are the wishes and efforts of man! How short his
> time, and consequently how poor will be his results, compared
> with those accumulated by Nature during whole geological
> periods! Can we wonder, then, that Nature's productions
> should be far "truer" in character than man's productions; that
> they should be infinitely better adapted to the most complex
> conditions of life, and should plainly bear the stamp of far
> higher workmanship? (Darwin p. 73).
> En quoi est-il moins admissible de chercher à faire une cellule
> que de chercher à faire une molécule? (Leduc Chap. 2).

**Abstract** The law answers to new technologies in three different ways: the technologies are enabled and restricted by regulation, they are promoted by (among others) granting intellectual property rights, and they may be required to share benefits. All of these functions come into play in synthetic biology. They are discussed in the present contribution. Concerning the enabling and regulating function the study suggests that the existing regulatory framework for genetically modified organisms (GMOS) cannot be relied on as an adequate means of controlling risks from synthetic biology. Various strands of synthetic biology are either not captured by the present regulation, or not appropriately treated by the present risk assessment methodology. This study suggests that based on a screening and categorisation of the risks from synthetic biology new regulation should be introduced covering both genetic engineering and synthetic biology. Concerning the promotion of the technology through intellectual property law the article finds that patent law has primarily served interests of industry to privatise nature. This has hindered the pluralistic

This article has very much benefitted from the discussions in the working group of the European Academy. I am particularly grateful to Michael Bölker, Margret Engelhard and Broder Breckling for many helpful suggestions. Part of this article was published in B. Giese, Chr. Pade, H. Wigger, A. von Gleich (eds.) Synthetic Biology - Character and Impact, Springer, Berlin, 2014.

G. Winter (✉)
Universität Bremen, Fachbereich Rechtswissenschaft, 28353 Bremen, Germany
e-mail: gwinter@uni-bremen.de

M. Engelhard (ed.), *Synthetic Biology Analysed*, Ethics of Science
and Technology Assessment 44, DOI 10.1007/978-3-319-25145-5_7

171

evolution of research and development (R&D). Besides advocating common pool solutions, the article explains how the preconditions of patenting can be interpreted in a way that sparks free R&D in synthetic biology. Concerning benefit sharing the new legal regime on access and benefit sharing is applicable also to synthetic biology. The increasing artificiality of its products however poses the question how the contribution of the original genetic resource can be tracked and valued. It is suggested that rather than relying on the bilateral exchange 'genetic resource for remuneration' joint R&D projects between provider and user should be strived for.

## 7.1 Introduction

Synthetic biology comprises a variety of modifications and constructions of life forms, all gradually replacing natural with artificial design and/or material. There is no clear-cut definition of synthetic biology so far, but it exhibits features that go far beyond genetic engineering and other biotechnologies. Three main strands of synthetic biology research and development (R&D) may be distinguished [1]:

- Genetic modification of organisms at larger scale (radical genetic engineering): It envisions the assembly in organisms of novel genomes from genetic parts (i.e. "biobricks" or "bioparts") and devices (functional units). These bioparts may be
    - natural genomic sequences that are being applied to a new purpose,
    - natural genomic sequences that have been redesigned to function more effectively, or
    - artificial genomic sequences that have been designed and synthesized from scratch and do not have any natural counterpart.

This approach includes the radical modification of existing cells, including supporting operations such as the construction of minimal cells (top down approach) that are reduced to a set of minimal functions and serve as chassis cells for further constructions (Danchin, forthcoming).

- The construction of standardized bioparts and of protocells by "bottom up" in vitro chemical synthesis. Protocells are vesicles that can exhibit certain organismic functions as for example growth, division or metabolism.
- Xenobiology, that is the construction of cells based on a chemistry not found in nature, e.g. based on nucleic acids or amino acids (so called Xenonucleic acid (XNA) and Xenoaminoacids) that are fundamentally different from (or "orthogonal" to) the molecular design of existing living organisms.

While the second and third strands of development are particularly unique, the first basically is still a kind of genetic modification. However, the modifications become more and more radical so that the resulting systems may have new features not found in existing natural living systems. Future organisms may reach the stage of complete synthesis and new assembly.

[1]See Chap. 1. in this volume.

In terms of cell functions the major lines of research concentrate on:

- novel cells as metabolic machines: they immensely enhance production technologies and product yield
- novel cells as biosensors[2]: whole cells that indicate the presence of a target analyte, as for example arsenic
- novel cells used as secure information storage device
- novel cells as experimental artefacts: made for scientific purposes, for the pleasure of experimentation, or for fun.

Given the early stage of synthetic biology R&D, the potential applications of the technology are not yet clear. They are envisaged to include computer technology, medicine, energy, fine chemicals, food, materials, environmental engineering and agriculture.[3]

Concerning potential benefits, there is hardly any comparative research. In some cases synthetic biology might be the only means to produce a certain end-product, while in other cases it may be a more efficient tool when compared to conventional genetic engineering. Until now only few examples exist. This is a shortcoming which makes it difficult to properly determine whether synthetic biology will generate benefits at all—and what kinds.

The risks of the technology are likewise widely unknown.[4] The closer synthetic biology comes to methods of conventional genetic engineering, the more knowledge—scarce as it still is—is available and the better risk assessment methodology developed for genetic modification—imperfect as it still is—can be used as a blueprint. But the more radical the technology becomes, the more uncertainty prevails. It may be that the situation is not yet pressing because the technology is in most cases still practiced in closed systems. But it should be ensured that it remains there (and that the containment itself is safe) and any deliberate release is prohibited before sufficient risk related knowledge is available.[5]

Besides safety issues, biosecurity must also be examined because synthetic biology may open up new potential for warfare.[6]

The law answers to new technologies in different ways, which can be divided into three functions:

- enabling and regulating: this somewhat paradoxical function of the law is characteristic for liberal societies. R&D is on the one side enabled by constitutional freedoms and certain legal infrastructure; on the other hand it is restricted by regulation based on countervailing constitutional guarantees

---

[2]Described by French et al. (2015) to be "a hybrid device consisting of a biological component, which provides the desired sensitivity and specificity, and an electrical transduction system, which converts the response to an output which can be monitored, analysed and recorded."

[3]Baldwin (2012) Chap. 7; Church and Regis (2012).

[4]For recent attempts to describe and explain risks of synthetic biology see Bölker (2015), Breckling and Schmidt (2015), Giese and von Gleich (2015).

[5]For a similar proposal see Friends of the Earth (2012).

[6]For an in-depth analysis in relation to the US see Maurer (2010/2011).

- promoting: governments provide incentives by various means, including R&D funding and the rewarding of R&D results by intellectual property rights
- redistributing: the law demands that those who draw benefits from R&D share them with other contributors.

In synthetic biology R&D all of these functions come into play and will be treated in this contribution. The crucial legal device enabling this technology is the constitutional freedoms of research and entrepreneurial activity. Concerning regulatory functions, existing laws ensuring biosafety and biosecurity must be checked whether they sufficiently control synthetic biology or must be amended by new laws and other standards of control (Sect. 7.2). Laws promoting the development of synthetic biology include a broad range of measures, of which R&D funding and intellectual property rights are the most important ones in the present context. The questions will be in what directions the funding programmes and intellectual property law guide synthetic biology R&D (Sect. 7.3). Finally, in relation to the question of redistributing, the new regime on access to genetic resources and benefit sharing (ABS) between providers and users of such resources must be considered.[7] Although most of the organisms presently used for synthetic biology (such as *Escherichia coli*) exist in countries which do not claim ABS rights, this situation may change in the future. An example is extremophile microorganisms from tropical countries, which may be used as source or recipient of synthetic biology constructs (Sect. 7.4).

## 7.2 Enabling and Regulating Synthetic Biology in View of Environmental and Health Risks

As synthetic biology is a new technology, no specific regulation has as yet been introduced. This raises the following questions: whether constitutional law, while enabling synthetic biology, also allows or even requires that a regulatory framework must be introduced (Sect. 7.2.1), what parts of the technology are already covered by existing legislation (Sects. 7.2.2 and 7.2.3), and in what respects the current risk assessment and management approach must be adjusted to the specifics of synthetic biology (Sect. 7.2.4).

### 7.2.1 Constitutional Freedoms and the Regulation of Synthetic Biology

The current dynamic evolution of synthetic biology research is enabled by the constitutional freedom of scientific research. This freedom has been laid down in Art. 13 of the EU Charta of Fundamental Rights (ChFR) and, in Germany, by Art. 5 para 3

---

[7]For an overview see Kamau and Winter (2015).

Grundgesetz (GG).[8] Both provisions, however, allow for regulatory restrictions if public interests so require. While these interests can be politically defined by the legislator at the EU level,[9] they need to have a basis in competing basic rights or principles under the German Grundgesetz (GG),[10] such as the right to health and the obligation to protect the environment.[11]

Likewise, economic activities in synthetic biology are enabled by constitutional rights, namely the freedom of professions and the freedom of enterprise.[12] But also these freedoms can be restricted in the public interest by regulation.[13]

While the legislator has wide powers to introduce regulatory measures, it may even be obliged to do so. As the legislator is bound to apply the precautionary principle,[14] full knowledge about the size and likelihood of harmful effects is not required for the adoption of regulatory instruments. On the other hand, however, they must not be based on pure speculation. There must be an indication of a risk to human health or the environment. In relation to synthetic biology, as mentioned above, risk-related knowledge is widely lacking with regard to some of its recent developments, such as xenobiology and protocell research and engineering. But interpolations may be possible from better known research. To the extent they reveal indications of risk, the legislator is empowered and even obliged to regulate, such as by introducing prior risk assessment and approval.

German constitutional law provides one further basis for an obligation of the state to legislate on synthetic biology, which shall be sketched out here because it might also serve as a model for EU constitutionalism. This basis is the so-called essentiality principle (Wesentlichkeitstheorie). The principle addresses the question under what circumstances a parliamentary law must be introduced, and how exhaustively it must rule on an issue. Traditionally, the problem at stake was the separation of powers between the legislatory and executive branch. The principle of "legal reservation" (Vorbehalt des Gesetzes) was developed to ensure democratic legitimation of executive action. It stipulates that executive action needs a basis in a parliamentary law if it intrudes into the basic freedoms of individual citizens. This principle was later extended by jurisprudence of the Federal Constitutional Court to

---

[8]There is no specific freedom of research in the European Convention on Human Rights (ECHR) but research is considered to be part of the freedom of opinion according to Art. 10 ECHR.

[9]See Art. 52 para 1 ChFR.

[10]See Art. 5 para 3 sentence 1 GG which guarantees the freedom without any restriction. Jurisprudence has however established that restrictions are possible and even mandatory if based on other constitutional provisions.

[11]Art. 2 para 2 sentence 2 GG, Art. 20a GG.

[12]Art. 15 and 16 ChFR. In the German Grundgesetz both of these freedoms are derived from Art. 12.

[13]Art. 52 para 1 ChFR; Art. 12 para 1 sentence 2 GG.

[14]Art. 192 TFEU. The German Grundgesetz includes an obligation of the state to protect human health on the basis of Art. 2 para 2 sentence 1. This article is interpreted as involving the precautionary principle (BVerfGE 49, 53 (Kalkar). Concerning environmental protection an obligation to protect is based on Art. 20a. However, this clause cannot be read as requiring precautionary measures. It is thus less "environmentalist" than Art. 192 TFEU.

also cover those executive activities which do not directly intrude on basic freedoms but indirectly influence the enjoyment of the same, or, more generally, touch upon "fundamental normative areas" ("grundlegende normative Bereiche").[15] It has since been called the "Wesentlichkeitstheorie" (essentiality principle). In the leading decision the court applied the principle to the fast breeder technology.[16] This decision could be understood to mean that not only state action but also private activities, if of crucial importance for society, must be based on a law. Apart from nuclear energy, the question was also posed in relation to other new technologies[17] such as genetic engineering,[18] non-ionising radiation from mobile radio,[19] nanotechnology, and agroindustrial techniques such as the keeping of laying hens in cages.[20] It is true that the courts which supported this extension found, in the normal case, that the new technology was already sufficiently captured by existing legislation. But this was founded on their basic conviction that the Wesentlichkeitstheorie was applicable not only to state measures but also to societal activities. The ruling opinion of scholars, however, opposed this, arguing that the principle of a free society would be reversed if a new technology were allowed only if based on a legal act.

My opinion is that the Wesentlichkeitstheorie should, indeed, be understood as a request for a parliamentary legal basis for new technologies that may endanger basic rights and the environment, even (or especially) if knowledge about its effects is not yet conclusive. Such law would certainly require that synthetic biology, be it exerted by state-based institutions or by private actors, must be subject to risk assessment.[21] In order not to hinder scientific progress, the law could introduce different degrees of control depending on a preliminary estimate of the hazard potential of different lines of research. If there is sufficient knowledge to predict that certain categories of research will not cause harm, those categories can even be freed from further risk assessment.[22]

---

[15]BVerfGE 49, 53 (126) (Kalkar).

[16]BVerfGE 49, 53 (127) (Kalkar).

[17]Further policy areas where the principle was invoked include such diverse issues as the protection of young people from pornography (BVerfGE 83, 130 (142)), the governmental subsidisation of the media (BVerfGE 80, 124 (132)), the deployment of armed forces for military purposes (BVerfG 90, 286 (383)). In still other cases the court applied the Wesentlichkeitstheorie to imbalances of powerful and powerless positions which needed reequilibration, such as concerning the rights of trade agents after dismissal by their principal (BVerfGE 81, 242, 254 et seq.) and the legal hedging of struggles between capital and labour (BVerfGE 84, 212, 226 et seq.).

[18]VGH Kassel, judgement of 6 November 1989, NJW 1990, 336 = NVwZ 1990, 276, available in German at http://openjur.de/u/289489.html (accessed 27.07.2015); see further below Footnote 39.

[19]OVG Lüneburg, Decision of 06.12.1993-6 M 4691/93, DVBl. 1994, 297.

[20]BVerfGE 101, 1 (34 et seq.).

[21]In addition, an essential impact on society could also be seen in synthetic biology's challenge of moral values. We will come back to this in Chap. 6.

[22]Such preliminary identification of the parts needing further risk assessment is a common technique of environmental law, as for instance in the law on environmental impact assessment. It has also been proposed for nanotechnology by the German Expert Commission on Nanotechnology.

The envisaged law might even take a much broader approach and be developed as a comprehensive basis for all kinds of biotechnology, including modern breeding techniques, traditional genetic engineering and synthetic biology. This would overcome the present inconsistencies, which introduce a very strict regulatory regime for certain kinds of biotechnology (such as genetic engineering) while others are largely left untouched (such as certain kinds of highly intrusive breeding), although the latter might be more consequential for biodiversity and human health[23].

Insofar as any existing legislation already covers potential impacts of synthetic biology on human health and the environment, and appropriately so, no new law addressing synthetic biology is constitutionally required. In this vein we need to check what risk regulation is applicable to synthetic biology, and whether it covers the relevant risks posed by this technology.

The regulatory framework that must be scrutinised for this purpose comprises preventive administrative law and ex post civil liability. We will discuss them in turn.

## 7.2.2 Preventive Administrative Regulation

Preventive administrative regulation means that actors in the field of research, development, production, trade and use of synthetic biology are subjected to a set of duties of care concerning effects on third parties or public goods. The fulfilment of these duties is supervised by administrative bodies. Third parties may be given rights to claim protection against risks. To grasp the whole regime one speaks of an administrative law relationship between actors, affected third parties and administrative bodies.

Most closely related to synthetic biology is the legal regime on genetically modified organisms (GMOs). Other regimes (which are not considered here) are the regulation of chemicals and that of pathogens. The GMO regime consists of both EU and MS legislation. It is basically structured according to whether GMOs are handled in containment, or intentionally introduced into the environment, be it through release at a predetermined site or, after they have been placed on the market, through introduction anywhere.

In the EU any works or products based on genetic modification are subjected to a special legal regime for GMOs. The EU thus hooks its regulation up with the technology. In contrast, in the US processes and products are checked as part of the control regime for non-modifying processes and non-modified products.

It has been debated what concept is more appropriate. The proponents of the US approach allege that the EU overregulates the issue. They regard many kinds of genetic engineering as safe and criticise that in the EU two authorizations may

---

[23]See further G. Winter, P. Knoepfel, H.-P. Fricker, The biotechnological utilisation of genetic resources and its regulation, Biel: Durabilitas 2014 (http://www.sanudurabilitas.ch/uploads/downloads/5/Durabilitas_2014_Genetic_resources.pdf)

be needed for one product, such as in the case of a genetically modified pesticide plant that would require one authorization under GMO and the other under pesticides legislation while in the US just one authorization under the pesticides legislation would suffice (Lynch and Vogel 2001). Vice versa, the proponents of the EU approach doubt that in the US the risks are adequately assessed. Concerning products, while in the EU all genetically modified products are subject to administrative oversight, this is in the US only the case for a limited range of products which are considered to be intrinsically dangerous for other reasons than their being genetically engineered, such as drugs, toxic substances, foods, seeds, fertilizers, and pesticides. Other genetically modified products which fall outside that scope are however imaginable, such as microorganisms engineered for soil decontamination or for energy production. Synthetic biology might contribute even more, such as products for purposes of construction, computing, leisure, and arts. The product-related US approach will therefore need to be extended to new products. In addition, pre-market authorization may have to be introduced for some product lines. With that, a shift of the burden of proof of safety which currently widely rests with the authorities might have to be shifted to the producers.[24]

Before considering whether the EU's GMO regime is an appropriate regulatory tool for synthetic biology we need to examine if, and to what extent, the existing EU GMO regulatory regime is applicable to synthetic biology at all.

### 7.2.2.1 Applicability to Synthetic Biology of the GMO Regime

The GMO regime is, as already mentioned, applicable to the "contained use," the individual "release" and the wider "placing on the market" of "GMOs." The regulation of contained use is harmonized EU wide only in relation to genetically modified microorganisms (GMMs).[25] Contained use of other GMOs is thus left to the regulatory competence of the Member States (MS).[26] In contrast, the regulation of the release and placing on the market of GMOs is standardized by EU legislation concerning all kinds of GMOs.[27] In any case, the core notion triggering the regulatory regime is a GMO. Its legal definition must therefore be explained and applied to synthetic biology techniques. The legal definition varies to some degree in relation to GMOs in general and GMMs, but the differences are not important in the present context.

---

[24]See further Paradise and Fitzpatrick (2012/2013) and Mandel and Marchant (2014/2015).

[25]Art. 1 Directive 2009/41/EC on contained use of GMOs.

[26]The German Act on Gene Technology (Gentechnikgesetz—GenTG), for instance, extends its provisions on contained use to all GMOs. It however empowers the government to exempt those GMOs which are considered to be safe (Sect. 2a GentG).

[27]Art. 2 (1) Directive 2001/18/EC on the deliberate release into the environment of genetically modified organisms.

An "organism" is legally defined as

any biological entity capable of replication or of transferring genetic material.[28]

This already excludes from the application of the GMO regime any modified or artificial subcellular bioparts that are not capable of replication.
Further, a genetically modified organism is defined as

...an organism, with the exception of human beings, in which the genetic material has been altered in a way that does not occur naturally by mating and/or natural recombination.[29]

Thus, for a GMO an organism must exist that is modified in certain artificial ways. For synthetic biology, this means that the GMO-regime only deals with activities which start with a real organism and modify it in specified ways. This excludes from the regime the complete synthesis of a known organism as well as the completely new design and synthesis of a new organism. In particular, bottom-up constructed protocells are not covered by the GMO regime.

The third element of the definition of a GMO is that the "genetic material" of the organism has been altered. The term "genetic material" undoubtedly includes the DNA and arguably also the RNA, considering the fact that the mRNA and tRNA, switched on by a gene, are part of the information process initiating the production of amino acids and through them of proteins. However, if by methods of the so-called xenobiochemistry (Budisa 2012) the amino acids are replaced by non-natural ones, and thus, new proteins emerge creating hitherto unknown properties of the organism, the "genetic material" is not altered but rather the higher level of derived chemicals.

The fourth element is that the genetic material contained in the organism was "altered". This poses the question if "alteration" also includes the complete replacement of the genome of a cell, such as in the experiment with mycoplasma bacteria of the Craig Venter Institute (Gibson et al. 2010). Based on a teleological reading this would, because of the unknown risks, need to be controlled even more than the mere modification. However, in a literal interpretation the full replacement is different from a mere alteration. One may ask if this should be different in the case in which the inserted material consists of newly synthesized conventional components. But in this case the organism is not altered but remains the same both chemically and functionally.

The fifth element is that the nucleic acid molecules inserted into a host organism may have been "produced by whatever means outside an organism."[30] This means that traits taken from an existing other organism are covered, but also a synthesized copy of them, and even synthesized traits having a new design, such as

---

[28]Art. 2 (1) Directive 2001/18/EC.

[29]Art. 2 (2) Directive 2001/18/EC.

[30]Art. 2 (2) together with Annex I A Part I (1) Directive 2001/18/EC.

those generated by xenobiology (Schmidt 2010; Budisa 2012).[31] This means that xenobiology insofar it induces artificial DNA or RNA is included in the GMO regime.

The sixth element is that nucleic acid molecules must be inserted into a host organism. This excludes from the GMO-regime methods of reducing organisms to minimal cells because in this case genetic material is removed from, rather than added to, the organism.[32]

The seventh element, as mentioned, is that the alteration of the genetic material is done "in a way that does not occur naturally by mating and/or natural recombination".[33] The core techniques qualifying as not natural are listed in Annexes to the relevant directives. They include, inter alia, the insertion of nucleic acid molecules by means of a vector system into a host organism in which they do not naturally occur, or by direct introduction such as micro-injection, or by not naturally occurring cell fusion or hybridisation.[34] This implies, for instance, that the gene gun method used in the do-it-yourself networks (DIY-Bio) is un-supervised.[35]

In contrast to the positive list of techniques qualifying as genetic engineering, certain techniques are excluded from the GMO regime because although being more or less artificial they can (at least theoretically) also occur under natural conditions. These techniques are mutagenesis and certain kinds of cell fusion.[36] However, a whole bunch of "New Plant Breeding Techniques (NPBTs)"—arguably included in a broad understanding of synthetic biology—have been argued that although in principle "natural" are so deeply interfering that they can be as hazardous as GMOs in the legal sense. Such techniques include targeted site-specific mutagenesis, transgenesis as an intermediate step of breeding processes where the transgene is subsequently removed, or "cisgenesis" where genes from the same species or family are transferred (Parisi 2012; Raaijmakers 2009). Thus, a substantial part of new breeding techniques appear not to be captured by the EU GMO regime (Table 7.1).

In conclusion, synthetic biology, insofar as it works on existing living cells and alters their genetic material in a way that does not occur naturally, must be counted as a technique resulting in genetic modification and thus as subjected to the existing EU GMO regime. In particular, organisms in which the genetic content was modified by synthesized material of natural or artificial design are

---

[31]This technique was however not unknown to earlier genetic engineering. For instance, the gene which encodes the PAT-protein and conveys tolerance of the herbicide glyphosate was redesigned and thus differs from the natural PAT-gene. Example taken from (Bundesregierung 2011). The radical version would be the above cited mycoplasma experiment.

[32]For a description of this technique see Budisa (2012), pp. 103–108.

[33]Art. 2 (2) Directive 2001/18/EC.

[34]Directive 2001/18/EC Art. 2 (2) together with Annex I A Part I (1)–(3).

[35]How naïvely the networks operate can be studied from the video displayed at http://www.sueddeutsche.de/wissen/biohacking-bewegung-leuchtende-pflanzen-zum-selberbasteln-1.1875586-2 (visited 27.07.2015).

[36]Directive 2001/18/EC Annex I B.

**Table 7.1**  Defining GMOs in application to synthetic biology

| Elements covered by the legal definition of a GMO | Elements not covered |
|---|---|
| The GMO must be an organism | Bioparts |
| The GMO must derive from an organism | Complete synthesis of an organism; bottom-up construction of a protocell |
| The <u>genetic material</u> must be altered | Modification on level of amino-acids or proteins (xenobiochemistry) |
| The genetic material must be <u>altered</u> | Complete replacement of the cell content, be it with conventional or new design |
| The inserted transgenes can be of any design and construction method | – |
| Transgenes must be "inserted" | A minimal cell |
| Positive and negative lists of techniques of insertion | Not listed techniques (e.g. gene gun), new breeding techniques |

covered, even insofar as new genetic xeno-material is introduced. By contrast, the following synthetic biology products are not captured by the GMO regime:

- an organism which was synthesized, be it of natural or artificial design
- an organism in which the genetic material was completely replaced by known or artificial genetic material
- an organism into which genetic material was inserted by other techniques than vector systems, micro-injection, non-natural cell-fusion or hybridization
- an organism whose chemical derivatives (amino acids, proteins) were modified
- a protocell
- a minimal cell
- synthesized or extracted bioparts
- an organism resulting from new breeding techniques which although in principle naturally occurring are deeply interfering.

It appears that this result—important synthetic biology techniques not being covered by the GMO regime—is not adequately discerned by research institutions and governments.[37]

There are two ways of reacting to the fact that parts of synthetic biology escape the scope of the existing GMO regime: One is to widen the scope so that more areas of synthetic biology are covered, and the second is to introduce a new law. The first option is certainly easier to reach politically, but the second would be more appropriate because it could be based on a new approach which better matches instruments of administrative oversight with different categories of risk. This approach could even release those kinds or results of genetic engineering that can be considered safe from regulation or at least from a full premarket control.

---

[37]See for Germany Acatech et al. (2009), p. 34. Bundesregierung (2011).

### 7.2.2.2 Adequacy of the GMO Regime

**Introduction**

We now proceed to consider what principles of risk assessment are appropriate for synthetic biology. This shall be done by critically reviewing the risk assessment methodology that is presently applicable to GMOs. If they are found to be inappropriate, better methodologies for synthetic biology must be introduced.

Products from synthetic biology, or synthetic biology products (SBPs)[38] will probably be fabricated and used in contained systems for a long time to come. Therefore, the relevant EU legal acts on contained genetic engineering operations must be consulted for their adequacy for SBPs. However, it is also possible that SBPs will be developed that shall intentionally be introduced into the environment, such as microorganisms for the treatment of contaminated water or soil; or for the production of energy from biomass (French et al. 2014). It is less probable that SBPs will be placed on the market for random release in the near future. But the possibility exists, for instance for microorganisms constructed for environmental management or energy fabrication. It must also be considered that a vibrant market has emerged for bioparts which provides services for contained R&D in synthetic biology. We will therefore first explore the regime for contained operations, and then the deliberate release of SBPs at certain locations as well as their market placement.

**Contained Use**

As already indicated, EU law on contained use of GMOs only refers to genetically modified microorganisms (GMMs) leaving other GMOs to the legislative competence of the member states.

*Risk Paths*

Even if kept in containment, GMMs may cause risks for the researchers and workers. Moreover, they may unintentionally leak into the environment through persons carrying them out of the lab, or through solid waste, sewage or exhaust disposed from the lab. The same paths must be considered for SBPs.

*Protected Goods*

According to Art. 4 Directive 2009/41/EC Member States

> ...shall ensure that all appropriate measures are taken to avoid adverse effects on human health and the environment which might arise from the contained use of GMMs.

The goods protected by the GMM regime are thus human health and the environment. Any "adverse effect" to them must be avoided.

Although this is not explicitly mentioned in the directive it has been discussed whether besides preventing risks GMMs must also provide a socio-economic

---

[38]I suggest this term for the emerging debate on a regulatory scheme for synthetic biology. Alternatively one could consider "SynBio organism", but this would not cover bioparts.

benefit. When in the late eighties and early nineties the first facilities with contained systems were built for research on dangerous microorganisms, concerns were raised if the containment would be perfect enough to hinder any escape of GMMs. Considering that a residual risk of leakage cannot be avoided, it was debated if the unavoidable remaining risk should not be weighed against the benefits generated by the GMM. For instance, in a hearing on the construction of a BASF facility for the production of the pharmaprotein Tumor Necrosis Factor (TNF) a concerned citizen argued that TNF was ineffective if not detrimental as a medicinal drug so that the construction of a production unit for TNF constituted, as she called it, a senseless risk (cf. Winter et al. 1993:34). Since then, the discussion about weighing risks against social benefits (or their absence) has faded away in relation to contained systems. It has however continued in relation to the deliberate release and market distribution of GMOs.[39]

Concerning highly problematic kinds of SBPs the same discussion may be reopened even in relation to contained systems.

### Burden of Submission of Risk Related Data

Risk assessment is only possible if appropriate data are available. Generally, in administrative proceedings the authorities are responsible for collecting the relevant data (investigation principle).[40] Ultimately, this rule rests on the fundamental right to individual freedom, which implies that if a law imposes restrictions based on certain factual circumstances these facts must be identified and proven by the competent authority.

The burden of producing evidence can however be imposed on the individual by special legislation. This normally occurs, if an activity requires prior authorization or notification, because it is assumed that the activity is suspected to pose a risk and shall therefore only be allowed after detailed examination. The EU GMM regime is based on this assumption and therefore shifts the burden of data provision to the applicant.[41] It specifies which data have to be presented, limiting the scope to those data which are needed to assess whether the substantive protective standard (the protection of human health and the environment) is met.[42]

If the presented data are not sufficient to allow a prognostic assessment, the competent authority can request the submission of additional data.[43] If the available knowledge is not sufficient for this purpose, the applicant bears the burden of generating it, provided there are indications of risk.[44]

---

[39]See further below.

[40]See Art. 337 TFEU and von Danwitz 2008, 417–421. For Germany see Sect. 24 Administrative Procedure Act (Verwaltungsverfahrensgesetz—VerwVfG).

[41]It is true, however, that Directive 2009/41/EC allows for exempting from its scope those GMMs which are considered to be safe (Art. 3 (1) (b) together with Annex II Part C of the same directive).

[42]Arts. 6–9 Directive 2009/41/EC.

[43]Art. 10 (3) (a) Directive 2009/41/EC.

[44]This requirement can be based on Art. 4 Directive 2009/41/EC as interpreted in view of the precautionary principle according to Art. 191 (2) (2) Treaty on the Functioning of the European Union (TFEU). On the necessity of indications and thus the exclusion of a zero risk approach see European Court, Case T-13/99, judgment of 11 September 2002 (Pfizer), paragraphs 144-148.

Knowledge relevant to an authorisation or notification proceeding may already be held by the administrative authority. If that is the case, the authority must make use of it in the authorisation procedure and cannot ask the applicant to reproduce it anew.[45]

It appears that these principles of data submission would also fit if an authorisation regime for using SBPs in contained systems was introduced.

### List of Data to be Submitted

In the case of contained use of highly hazardous GMMs the data to be submitted by the applicant comprise the following:

(a)  [...]
(b)  the recipient or parental micro-organism(s) to be used,
      the host-vector system(s) to be used (where applicable),
      the source(s) and intended function(s) of the genetic material(s) involved in the modification(s),
      the identity and characteristics of the GMM,
      the culture volumes to be used;
(c)  a description of the containment and other protective measures to be applied, including information about waste management, including the type and form of wastes to be generated, their treatment, final form and destination,
      the purpose of the contained use, including the expected results,
      a description of the parts of the installation;
(d)  information about accident prevention and emergency response plans, if any:
      any specific hazards arising from the location of the installation,
      the preventive measures applied, such as safety equipment, alarm systems and containment methods,
      the procedures and plans for verifying the continuing effectiveness of the containment measures,
      a description of information provided to workers,
      the information necessary for the competent authority to evaluate any emergency response plans, if required under Article 13(1);

While the data listed sub (c) and (d) might be transferable to the situation of hazardous SBPs those sub (b) reflect the fact that the object of assessment is genetic modification of existing organisms. This may be appropriate for SBPs that are based on existing organisms. However, for new SBPs lists of required data must be developed that are better targeted to the specific risks of such SBPs. Where interpolations from donor, vector and recipient organisms are not possible specific tests concerning the resulting organism must be required. Moreover, as the GMO regime only covers living organisms, risks from bioparts, individually and in combinations, are not addressed by the data list.

---

[45]See the clause "if necessary" in Art. 10 (3) Directive 2009/41/EC.

## Assessing and Categorising Risk and Containment

Risk prevention measures should differ depending on the severity of the risks caused. The more hazardous the use of an organism is the tighter the containment must be. This is also the logic applied in the EU GMM regime. Four risk categories are distinguished corresponding to an increasing intensity of containment measures. These categories are described as Class 1: no or negligible risk, Class 2: low risk; Class 3: moderate risk; and Class 4: high risk. The four risk classes are correlated with four containment classes. These consist in clusters of measures concerning the construction of the lab (e.g. isolation), the equipment (e.g. negative pressure), the system of work (e.g. restricted access, clothing), and the treatment of waste (e.g. inactivation of GMMs).[46]

The risk assessment serves to classify any use of GMMs into one of the four risk and containment classes. A two-step procedure is recommended for this exercise[47]:

Procedure 1

Identify potentially harmful properties (hazard) of the GMM and allocate the GMM to an initial class (class 1–class 4), taking into account the severity of the potentially harmful effects.
and
Assessment of possibility of harmful effects occurring by consideration of exposure (both human and environmental), taking into account the nature and scale of the work, with containment measures appropriate to the initial class allocated.

Procedure 2

Determination of final classification and containment measures required for the activity.

Confirm final classification and containment measures are adequate by revisiting Procedure 1.

When assessing the risk of the resulting GMO, the hazards of the donor as well as the resulting organism must be considered, i.a.[48]:

(i)   the recipient micro-organism;
(ii)  the genetic material inserted (originating from the donor organism);
(iii) the vector;
(iv)  the donor micro-organism (as long as the donor micro-organism is used during the operation);
(v)   the resulting GMM.

---

[46]Art. 4 (3) and Annex IV of Directive 2009/41/EC.
[47]Commission Decision 2000/608/EC, Annex Nr. 2.
[48]Annex III A (2) Directive 2009/41/EC.

The following endpoints must be examined[49]:

Human health considerations:

expected toxic or allergenic effects of the GMM and/or its metabolic products,
comparison of the modified micro-organism to the recipient or (where appropriate)
parental organism regarding pathogenicity,
expected capacity for colonisation,
if the micro-organism is pathogenic to humans who are immunocompetent,
diseases caused and mechanism of transmission including invasiveness and
virulence,
infective dose,
possible alteration of route of infection or tissue specificity,
possibility of survival outside of human host,
biological stability,
antibiotic-resistance patterns,
allergenicity,
toxigenicity,
availability of appropriate therapies and prophylactic measures.

Environmental considerations:

Ecosystems to which the micro-organism could be unintentionally released from
the contained use,
expected survivability, multiplication and extent of dissemination of the modified
micro-organism in the identified ecosystems,
anticipated result of interaction between the modified micro-organism and the
organisms or micro-organisms which might be exposed in case of unintentional
release into the environment,
known or predicted effects on plants and animals such as pathogenicity, toxicity,
allergenicity, vector for a pathogen, altered antibiotic-resistance patterns, altered
tropism or host specificity, colonisation,
known or predicted involvement in biogeochemical processes.

These parameters will have to be revisited in relation to SBPs. Based on accumu-
lated experience, lists of typical organisms and treatments have been compiled for
GMMs. However, concerning SBPs, it is questionable if the research activities can
already be categorized in a like manner. They are still very diverse, and risk related
knowledge is scarce. Moreover, the risk classes and containment measures mainly
refer to the hazards of the donor and receiver organisms. It appears that for the
more radical interventions of synthetic biology into the genome, genuine methods of assessment must be developed. This is all the more the case in relation to
bioparts, protocells and minimal cells. Obviously, more discussion with scientists
is needed in this regard.

---

[49]Commission Decision 2000/608/EC, Annex Nr. 3.2.5.

## Introducing SBPs into the environment and placing SBPs on the market[50]

As already indicated, EU legislation, and in particular Directive 2001/18/EC categorises the introduction of GMOs into the environment as the deliberate release at a particular site and the introduction into the environment at any site after GMOs have been placed on the market. Both the release and the placing on the market must be authorised.[51] An authorisation of market placement of a GMO implies the subsequent introduction into the environment at any location, unless the allowable locations are restricted by conditions of the authorisation.[52] Concerning genetically modified food and feed, including seeds, a special regime has been established which takes precedence over the general regime. This will however not be treated in this article because synthetic biology is still far from resulting in food or feed.[53]

We can treat the deliberate release and the market placement together because the risk prevention criteria and risk assessment methodologies are largely the same for both activities, with certain variations due to the larger geographical scope of introductions into the environment of GMOs that are authorised for market release.

### Risk Paths

According to Art. 4 (3) Directive 2001/18/EC Member States

> shall ensure that potential adverse effects on human health and the environment, which may occur directly or indirectly through gene transfer from GMOs to other organisms, are accurately assessed on a case-by-case basis.

Correspondingly, an environmental risk assessment (ERA) must evaluate risks "whether direct or indirect, immediate or delayed, which the deliberate release or the placing on the market of GMOs may pose [...]."[54]

The distinction between direct and indirect effects means that not only those adverse effects caused by GMOs in direct contact with endpoints (e.g. a human being, animal or plant absorbing a GMO) have to be prevented but also those which are mediated by intervening factors. Annex II of Directive 2001/18/EC defines indirect effects as referring to:

> effects on human health or the environment occurring through a causal chain of events, through mechanisms such as interactions with other organisms, transfer of genetic material, or changes in use or management.

On this basis one could differentiate indirect effects further into natural causal chains (horizontal and vertical gene transfer, food chain, etc.) and chains mediated by human practices (such as agricultural change in pesticide use and crop rotation, etc.).

---

[50]The following analysis is based on von Kries and Winter (2011).

[51]Articles 5 and 6; 13-15 Directive 2001/18/EC which provide differentiated procedures of notification, risk assessment, commenting and final decision.

[52]Parts B and C of Directive 2001/18/EC.

[53]Regulation (EC) No 1829/2003 on genetically modified food and feed.

[54]Art. 2 (No. 8) Directive 2001/18/EC.

Concerning the distinction between immediate and delayed effects, the Commission Guidance on the environmental risk assessment (Commission 2001) gives examples for delayed effects such as the GMO developing invasive behaviour several generations following its release.[55]

In addition to alerting the risk assessment to direct/indirect and immediate/delayed effects the ERA must also consider different environments exposed to the GMO[56]:

> For each adverse effect identified, the consequences for other organisms, populations, species or ecosystems exposed to the GMO have to be evaluated.
> Moreover, there may be a broad range of environmental characteristics (site-specific or regional-specific) to be taken into account. To support a case-by-case assessment, it may be useful to classify regional data by habitat area, reflecting aspects of the receiving environment relevant to GMOs (for example, botanical data on the occurrence of wild relatives of GMO plants in different agricultural or natural habitats of Europe).

This rather ambitious programme, relating to genetically modified plants, was further elaborated by Guidance of 2010 of the European Food Safety Agency (EFSA) (EFSA 2010). It concentrates on interactions of the plant on the levels of organisms and ecosystems.[57]

While this analytical framework looks comprehensive, a note of caution is, however, appropriate: The fate of the GMO in the various environments may prove to be too complex to be examined. This is particularly true if the GMOs introduced into the environment are microorganisms. It is telling that in that regard the pertinent EFSA Guidance somewhat wearily states as follows:

> Predicting impacts of GMMs and derived food or feed on complex ecosystems can be difficult due to continuous flux and spatial heterogeneities in ecosystems creating a myriad of potential microbial habitats in which interactions between GMMs and their products with the indigenous organisms and/or abiotic components can take place. It is recognised that an ERA cannot provide data of a GMM or its products, which would cover all potential environmental habitats and conditions. Consideration of environmental impact (damage) should, therefore, focus on environments in which exposure is most likely or in which, when relevant, viable GMMs could potentially proliferate.

### Protected Endpoints

#### Human Health and the Environment

EU law has established that for the deliberate release of GMOs as well as for contained use the protected goods shall be human health and the environment. These shall be kept safe from 'adverse effects.' 'All appropriate measures' must be taken to prevent these.[58]

---

[55]Guidance Notes on the Objective, Elements, General Principles and Methodology of the Environmental Risk Assessment Referred to in Annex ii to DIRECTIVE 2001/18/EC, OJ L 18, 07.11.2003, p. 32.

[56]Annex II Directive 2001/18 Sect 4.2.2 and Commission Guidance (above Footnote 55) Sect. 3 3rd hyphen.

[57]See further on a multilevel approach of risk assessment of GMOs and SBPs Breckling and Schmidt (2015).

[58]Art. 4 Directive 2001/18/EC.

What are adverse effects? Is the mere presence of a GMO outside the field of release, per se, to be considered as adverse effect? Prevailing court practice and doctrine negate this. They posit that the adverse effect must be a result of such presence, like the damaging of non-target species from an insecticide plant. The justification given is that the law only addresses the specific risks of genetic engineering, which shall be only health and environmental risks.[59]

Concerning SBPs this might be seen differently. It could be argued that given the early stage of R&D in this area and the radically artificial nature of synthetic biology, SBPs should not be allowed to spread at all. Any release would then have to be contained. Alternatively, if SBPs were constructed to only survive under artificial conditions, one could consider their safe release into the environment, because they would immediately die off there. However, this would not apply to organisms which are intended to survive and perform in the open environment.

*Socio-economic Benefits*

GMO releases may create benefits for the producer and consumer. Is this to be weighed against the risks to human health and the environment? Such an analysis is envisaged in the genetic engineering legislation of some countries.[60] It is, however, only scarcely present in European GMO legislation.[61]

When pursuing this request two brands of risk-benefit-consideration should be distinguished: a risk-tolerating variant which would allow any risk that is outweighed by benefits, and a risk-averse variant according to which only residual risks can be outweighed by benefits. It is submitted that only the second variant should be adopted.

Concerning the release of SBPs into the environment, socio-economic benefits should also be introduced as an additional requirement; but only after its risks where assessed and found minimal, not as a vehicle to outweigh significant risks by higher valued benefits.

---

[59]See for Germany Administrative Court (VG) Berlin, decision of 12.09.1995 - 14 A 255.95, in: Eberbach/Lange/Ronellenfitsch, Recht der Gentechnik und Biomedizin, Entscheidung Chap. 4 on Sect. 16 GenTG; VG Braunschweig, judgment of 12. 9.1995 - 14 A 255.95, No. 27.

[60]For Germany see Sect. 16 paras. 1 und 2 GentG, according to which "harmful effects on the protected goods listed in Sect. 1 No. 1 must not be incurred if unacceptable in view of the objective of the release". Unacceptability in view of the release objective can be understood as a kind of weighing risk versus benefits. German scholars tend to reject such interpretation, arguing that this would be incompatible with the relevant EU law. See also Art. 10 of the Norwegian Gene Technology Act: "In deciding whether or not to grant an application, considerable weight shall also be given to whether the deliberate release will be of benefit to society and is likely to promote sustainable development." This provision has, however, rarely been applied in practice (Spök 2010).

[61]See the rather enigmatic opening clause ("...other legitimate factors") in Arts. 7 and 19 Regulation (EC) 1829/2003.

*Cultural Factors*

The rejection of GMOs by the majority of the population in a number of countries can be explained by cultural factors. This scepticism is based on a conglomerate of concerns including extreme precaution, criticism against neglecting the evolutionary wisdom, doubts about whether the promised benefits are not already available from existing organisms, political will as well as ethical concerns and religious beliefs.[62] The cultural factor is not well represented in national and international law as a legitimate justification for trade restriction. For instance, it was not even considered in the resolution of the WTO panel on EC restrictions concerning the marketing of biotech products.[63]

The ECJ has shown understanding for the cultural factor in *Commission v Poland* but finally rejected it by splitting the issue into three parts: Insofar as extreme precaution was alleged, the Court said that this does not dispense from the normal standard applied in the EU; concerning the opponent political will it held that the MS must neglect it once an EU legal act has been adopted; and concerning ethical and religious beliefs it held that the strength and spread thereof was not sufficiently proven.[64]

It is submitted that the cultural factor should be given a more legitimate place in regulatory designs concerning synthetic biology.[65]

*Data to be Submitted*

A long list of data has been compiled that must be submitted for an application for release of GMOs. It comprises[66]:

Information relating to the GMO
Characteristics of (a) the donor, (b) the recipient, or (c) (where appropriate) parental organism(s)
Characteristics of the vector
Characteristics of the modified organism
Information relating to the conditions of release and the receiving environment
Information relating to the interactions between the GMOs and the environment
Information on monitoring, control, waste treatment and emergency response plans

---

[62]Such skepticism has normally been neglected by the ruling mechanistic approach since its onset. See for a the complete lack of comprehension of the difference between a molecule and a (living) cell the citation introductory to this paper by Leduc (1912). Leduc is said to have coined the term synthetic biology. Darwin (see his statement introductory to this paper) seems to be more aware of such concerns although being considered the herald of mechanistic biological thinking.

[63]See further G. Winter, Cultivation restrictions for genetically modified plants: On variety of risk governance in European and international trade law, 1/2016 EJRR (forthcoming)

[64]ECJ Case 165/08, judgment of 16 July 2009 (Commission v Poland) paragraphs 54, 55, 58, 59.

[65]See further chapters of Pardo Avellaneda and Kristin Hagen in this volume.

[66]Annex III of Directive 2001/18/EC.

This list would have to be thoroughly checked for its suitability for SBPs releases. Once again, it must be considered that more and more research is aiming at replacing traits from parental organisms by synthesis and, even more importantly, by artificial design.

The ERA, as outlined by Annex II Directive 2001/18/EC, focuses on those paths of risk with human health and the environment as endpoints. Other endpoints, like the coexistence with non-GM agriculture, the economic benefit and political as well as cultural values, are hardly considered (Dolezel et al. 2009: 27). However, should these aspects become a legally required part of the risk management, then information has to be provided and assessed which is methodologically clear and rich in substance.

### The Stepwise Generation of Knowledge

Towards the end of the nineteen-eighties, when the deliberate release of GMOs was approached, knowledge about the involved risks was still highly undeveloped. Even today, there remain gaps in our knowledge. Nonetheless, to enable the release of GMOs and acquire knowledge, the step-by-step principle was introduced which stipulated the incremental generation of knowledge in parallel with decreasing containment of tests.

The step-by-step principle is characterised by recitals (24) and (25) Directive 2001/18/EC as follows:

> The introduction of GMOs into the environment should be carried out according to the "step-by-step" principle. This means that the containment of GMOs is reduced and the scale of release increased gradually, step by step, but only if evaluation of the earlier steps in terms of protection of human health and the environment indicates that the next step can be taken.
> No GMOs, as or in products, intended for deliberate release are to be considered for placing on the market without first having been subjected to satisfactory field testing at the research and development stage in ecosystems which could be affected by their use.

The following sequence of steps has emerged in practice:

laboratory
greenhouse
small-scale release into the environment under containment (not specified in law)
point release into the environment
placing on the market
subsequent random introduction into the environment

The substance of the step-by-step principle was somewhat specified by Commission Guidance which says that "data from each step should be collected as early as possible during the procedure." It points to the possibility that "simulated environmental conditions in a contained system could give results of relevance to deliberate release," such as the simulation of behaviour of microorganisms in the laboratory, and of plants in greenhouses.[67]

---

[67]Commission Decision 2002/623/EC, Chap. 2.

The step-by-step principle is an instrument of societal learning. In the initial phase of European genetic engineering legislation, it was at the fore of public debate and became a legal requirement as outlined.[68] With the amendment through Directive 2001/18/EC, post-release monitoring has become an additional instrument. In order to increase safety, and at the same time facilitate the release and market distribution of GMOs, it was emphasized that those issues which, for reasons of time or scale, cannot be solved at one level can be clarified through monitoring at the next level. Monitoring can therefore be seen as a phase of learning following the release or market distribution, respectively. This concerns especially the investigation of effects which cannot be researched on an experimental basis, such as complex interactions on population and ecosystem levels, or cumulative and long-term effects.

As to procedural aspects, the applicant must submit a monitoring plan that contains the following information[69]:

1. methods for tracing the GMOs, and for monitoring their effects;
2. specificity (to identify the GMOs, and to distinguish them from the donor, recipient or, where appropriate, the parental organisms), sensitivity and reliability of the monitoring techniques;
3. techniques for detecting transfer of the donated genetic material to other organisms;
4. duration and frequency of the monitoring.

The monitoring programme is then determined as a condition for the release authorisation. The operator is responsible for implementing the programme and reporting results to the authority.

It is submitted that the step-by-step-principle, including self-monitoring, should also be used in relation to synthetic biology. Of course, the methodology must still be adapted to the various strands of synthetic biology and its peculiarities.

### Steps in the Analysis and Assessment of Risks

It is characteristic for the risk assessment in form of the environmental risk assessment (ERA) that it processes the data successively in pre-defined steps. The staggered evaluation of risks is finally followed by the risk management, which translates the scientifically informed risk evaluation into measures, i.e. the authorisation, the conditions for the authorisation and, if applicable, the rejection of authorisation.

According to Annex II of Directive 2001/18/EC and the respective Commission Guidance the ERA consists of six steps. Using the language of the Annex the steps can be summarized as follows:

---

[68]The step-by-step procedure goes back to OECD reports, including OECD, Safety considerations for biotechnology, 1992.

[69]Art. 6 (2) (V) and Annex III C Directive 2001/18/EC.

In step 1, the inherent characteristics of the GMO are to be identified. They present factors (or "hazards") that can lead to risks depending on environmental conditions and usage.

In step 2, the potential consequences of each established adverse effect have to be evaluated. The evaluation concerns organisms, populations, species and ecosystems interacting with the GMO. Particular emphasis is given to the expected magnitude of the consequences. The latter can depend on the genetic design, the established adverse effects, the number of released GMOs, the receiving environment, the manner of the release and the control measures taken as well as on a combination of all these factors.

In step 3, the likelihood of the occurrence of each identified potential adverse effect is to be evaluated; here, each effect is examined individually, taking into account the risk factors, the number of released GMOs, the likelihood and frequency of gene transfer, the receiving environment and the conditions of the release.

In step 4, the different magnitudes of consequences (high, moderate, low or negligible) of every risk factor are linked to the different degrees of their likelihood (high, moderate, low or negligible). In addition, the overall uncertainty for each identified risk has to be described, including assumptions and extrapolations made at previous levels in the ERA, different scientific assessments and viewpoints, and the uncertainties contained in each evaluation.

In step 5, management strategies for risks from the deliberate release (or marketing) of GMOs are to be developed. The risk management is to be designed in a way so that identified risks can be controlled and that uncertainties can be covered. Safeguarding measures (coated seeds, isolation distances, etc.) have to be proportionate to the levels of risk and uncertainty.

In step 6, the overall risk of the GMO is determined. This consists of a summary of all identified risks and uncertainties of the examined application, taking into account the magnitude and likelihood of the adverse effects as well as the previous release of other GMOs. The achieved risk reduction caused by the management measures must also be considered.

Core to this 6 step procedure is the distinction between inherent factors of a GMO, adverse effects of these factors through interactions on the levels of the organism, populations, species and ecosystems, the magnitude of each adverse effect, and its likelihood. In addition, the uncertainties of the assessment shall be described. Safeguarding measures shall also be taken into account. This sounds thorough and comprehensive but may not sufficiently reflect the fact that synthetic biology is too diverse and unstructured to allow for a standardisation of risk assessment. For instance, the fact that much of the produce of synthetic biology is claimed not to survive under real world conditions must be integrated into the methodology (Giese and von Gleich 2015). Likewise, the focus on organisms does not reflect possible risks from bioparts and minimal cells.

*Familiarity*

The major innovation needed in risk assessment for Synthetic biology will be that the familiarity principle must be modified and finally even abandoned, because the newly designed organisms are intentionally more and more alienated from the genome of existing organisms.

The status of the familiarity principle in the GMO risk assessment can be summarized as follows: Risks to human health and the environment can be caused by traits of the non-modified parental lines and of the genetic modification. The concept of familiarity (or—using about the same approach—comparison with similar organisms or substantial equivalence), which goes back to an OECD paper of 1993, suggests that only effects of the genetic modification should be assessed. This is reasonable; otherwise the applicant could be blamed for adverse effects that are already contained in the parental line. However, critiques have alleged that, by focusing on the modification, the concept of familiarity cuts the organism into pieces and disregards effects of the newly created organism as a whole. Rather than assuming firm knowledge of the unmodified organism, one should rather look for the unexpected, the unfamiliar in interactions between the existing cause-effect network and the newly introduced GM component (Breckling 2004: 52–59).

Asking what the law demands in this regard, it should first of all be noted that the concept of familiarity is not conveyed by the wording of the substantive standard expressed in Directive 2001/18/EC. Rather, Art. 4 (1) states comprehensively that the release and the placing on the market of the GMO must not cause any adverse effects. The annexed rules on the ERA, however, state that a comparison with non-modified organisms "will assist in identifying the particular potential adverse effects arising from the genetic modification."[70] The new EFSA Guidance of 2010 unwisely reinforces this approach by making the "comparative safety assessment" the core yardstick of risk assessment.[71]

Whether called comparative or not, the examination is not allowed in any case to imply that the transgene has to be considered in isolation. Unintended position effects and mutual reactions at all organismic levels are rather the consequence of genetic modifications and have to be considered to their full extent. Upon closer look this is also envisaged by the EFSA Guidance of (2010). Therefore, the Annex on ERA is still right to regard the comparative approach as a heuristic, rather than constitutive, tool of the risk assessment.

Concerning synthetic biology, however, even this heuristic function will lose ground with the growing alienation from parental lines of the new synthetic organisms. New methods of risk assessment must be developed. It is suggested that such methodology should start with risk-related analysis of the main strands of developments of this technology. Subcellular parts and protocells, for instance, do not pose a risk of replication and through that of risks attached to life forms, such as

---

[70]Annex II Directive 2001/18/EC, C.

[71]EFSA (2010) Sect. 2.1.

becoming dominant in ecosystems. Rather, they are to be evaluated in terms of criteria used for chemicals, such as toxic, carcinogen, mutagen and allergen properties, persistence and bioaccumulation, as well as exposure analysis. Xenobiology is claimed to be safe because resulting organisms can only survive under very artificial circumstances. However, this is not necessarily true, so that scenarios and tests must be developed to prove this assumption. In addition, criteria used for chemicals should be applied. The major challenge will be to develop methods for the vast and ever-expanding works of those kinds of genetic engineering which increase the degree of artificiality even more. Specific tests must be developed in order to identify risks. Specific risk abatement technology must also be developed.

As all this costs time and effort, it appears to be advisable to establish a moratorium for the release into the environment of synthetic biology organisms, as well as a moratorium for the placing on the market of such organisms insofar as this entails any release into the environment.

## 7.2.3  Regulation Ex Post

Regulation ex post makes an actor liable to remedy or compensate for damage he or she has caused. There are various legal bases for such liability, general ones and ones specifically created for GMO- related risks.

The general scheme is tort liability. It presupposes that damage was intentionally or negligently caused to human health or material assets by an operator. The burden of proof, in principle, lies with the victim. Tort liability seldom leads to convictions because the causation and negligence are difficult to prove.

More specific and promising from the victim's perspective is strict liability for GMOs which has been introduced by some countries including Germany. Article 32 of the German Genetic Engineering Act (Gentechnikgesetz- GentG) provides:

> Where any properties of an organism that result from genetic engineering operations cause the death of a person or injury to his/her health, or damage of property, the operator shall be obliged to give compensation for the damage ensuing therefrom.

No intention or negligence is required. The proof of causation is facilitated in two ways:

> Causation from genetic engineering operations is presumed if the damage was caused by genetically modified organisms. The burden of proof that this was not the case lies on the operator.[72]

If the victim brings a prima facie proof that the damage was caused from genetic engineering operations of an operator the operator must disclose information "about the type of and steps involved in the genetic engineering operations performed" by her.[73]

---

[72]Art. 34 GentG.
[73]Art. 35 GentG.

In addition, the liability does not only extend to the victim's own damage but also covers expenditure incurred by her for the restoration of damage to the environment. If, for instance, a bacterium which has been gene-coded for an infectious animal disease escapes from the laboratory and causes a disease to bees, the operator is liable to pay for the forgone fruit yield and for the restoration of the bee population.

Directive 2004/35/EC on environmental liability establishes a third basis for liability. The concept does not introduce an additional right of a victim against an operator, but empowers and obliges administrative authorities to intervene. This is possible, i.a., if any deliberate release into the environment, transport and market placement of genetically modified organisms causes environmental damage.[74] The administrative authority can order the operator to take remedial action. NGOs are given rights to sue the authority if it remains passive.

Overall, synthetic biology as far as it is subject to the GMO regime, faces rather strict liability rules. As the special rules all refer to GMOs, they do not apply to technologies or products outside this scope. For this reason it must be considered whether the liability should be extended to those parts of synthetic biology which do not consist of GMOs in the legal sense, i.e. completely new organisms, organisms whose genome was completely replaced, organisms into which transgenes were inserted by other techniques than those contained in the positive and negative lists, organisms modified by xenobiochemistry, protocells, minimal cells, and bioparts.

### 7.2.4 Conclusion

Other than official statements by governmental and scientific bodies assume[75] the existing regulatory framework cannot be relied on as an adequate means of controlling risks from synthetic biology. Various kinds of synthetic biology are either not captured by the present regulation, or not appropriately treated by the present risk assessment methodology. This study suggests that the risks from synthetic biology should carefully and systematically be examined. On such basis new regulation should be introduced. This could be done by extending the scope and improving the risk assessment of the existing regulation on genetically modified organisms, or by taking a new approach that addresses biotechnology in a broad sense, including GMOs, synthetic biology, new breeding techniques and possibly further variants.

---

[74]Art. 3 para 1 and Annex III Directive 2004/35/EC.
[75]See for Germany Acatech et al. (2009), p. 34; Bundesregierung (2011).

## 7.3   Promoting Synthetic Biology Through the Granting of Intellectual Property Rights

Intellectual property is generally considered to be a powerful incentive to engage in R&D.[76] As in genetic modification, patents are the kind of intellectual property right which researchers will most often seek also in synthetic biology. A count of present patents in the field reveals that the rate is still low, but a significant increase is expected with the further development of the technology towards applications. Most of the patents are related to knowledge generation, enabling technologies and engineering principles (van Doren et al. 2013).

I will start with some basic reflections on patent law in biotechnology and then turn to more concrete problems of applying the existing legal framework.

### 7.3.1   Patenting Life Forms

Patents are, in spite of the many international treaties concluded for harmonisation purposes, still national constructs. This means that the procedure of granting, the preconditions and content as well as the sanctioning mechanisms are all nationally regulated and administered and judicially supervised by national institutions. Only in Europe, among the parties of the European Patent Convention, has the harmonisation progressed towards a quasi European patent, which although being provided by a centralised institution, the European Patent Office, is however still constructed as a national right subject to national implementation mechanisms. In the EU the harmonisation has even more been intensified by joint standards concerning biotechnology.[77]

Content-wise there are, in spite of a wide-ranging harmonisation of standards and procedures, some significant differences between the European and the American concept of patenting. In principle the US system has been more patent friendly and the European more restrictive. Very stimulating for US biotechnology has been the Supreme Court decision of 1980 in *Diamond v Chakrabarty* with its now proverbial cite that Congress had intended patentable subject matter to "include anything under the sun that is made by man".[78] The Court stresses human ingenuity (such as genetically modifying a microorganism) as a reason for patenting, as opposed to products of nature. If human impact can be shown life forms are

---

[76]Subsidies are another tool of promoting a new technology. Subsidiy strategies will however not be discussed in the present paper.

[77]Directive 1998/44/EC.

[78]Diamond v. Chakrabarty, 447 U.S. 303 (1980), Chap. III. See for the intentions of the ruling majority of the judges to encourage biotechnology Holman (2015), pp. 404 et seq.

open to patenting, including plants and animals.[79] In contrast, Europe while being less reserved about patenting phenomena of nature has more exceptions when phenomena of nature consist of life forms. Exempted is therefore the patenting of animals, plant varieties, biological procedures and therapeutic procedures. Europe is also more inclined to make use of the ordre public-exception if life forms are patented (Mueller 2006). With some simplification the difference between the US and the EU may be characterised as a difference between Anglo-Saxon utilitarianism with its focus on the input of human craft and European ethics that acknowledge an intrinsic value of life.

In my own opinion, the reluctance to patent phenomena of nature in the US and life forms in the EU is most convincing if embedded in a sociological consideration: nature, and especially living nature belongs to humanity as a commons. There is no reason why an individual, simply because he or she adds some tiny property to the complexity of an organism, should have a hold on the whole evolutionary process which has elaborated the organism. Patenting natural processes is rather an economic strategy to privatise nature, or, in terms of materialist theory, a new method of ursprüngliche Akkumulation (original accumulation).

In spite of such conceptions of common property the drive towards patenting has proven to be more powerful. This is even the case with regard to the European exemptions concerning higher life forms. Whereas Article 53 (b) EPC excludes patents for animal varieties, the patenting of a particular animal specimen has been accepted: The EPO started this practice with granting a patent for the so-called Harvard onco mouse,[80] which had already received a US patent a few years earlier.[81] The EPO did not regard the patent claim which related to a 'transgenetic

---

[79]The patenting permissiveness of *Chakrabarty* was somewhat confined by two Supreme Court decisions of 2012 and 2013 which even more stressed the requirement of human ingenuity. In *Mayo* the Supreme Court excluded laws of nature from patenting (*Mayo Collaborative Services v Prometheus Labs.*, Inc, 132 S. Ct. 1289, 1296–1298 (2012)). The case concerned a test kit that helped to identify relationships between concentrations of certain metabolites in the blood and the likelihood that a dosage of a thiopurine drug will prove ineffective or cause harm. The court held that "the claims inform a relevant audience about certain laws of nature; any additional steps consist of well understood, routine, conventional activity already engaged in by the scientific community; and those steps, when viewed as a whole, add nothing significant beyond the sum of their parts taken separately." In *Myriad* the court denied the patentability of naturally occurring DNA (*Association for Molecular Pathology v Myriad Genetics, Inc.*, 133 S. Ct. 2107, 2111 (2013)). The case concerned two human genes mutations of which increase the risk of cancer. The claim was an exclusive right to the two natural genes as well as to synthesized genes that are identical with the natural ones but for the fact that only the exons but not the introns are synthesized. The court held that the natural genes are products of nature, but not the synthesized gene if deviating from the natural one. See Holman (2015): 409 et seq. for the legal uncertainty this jurisdiction has created with regard to the thousands of patents that had since *Chakrabarty* been granted on isolated DNA.

[80]See Harvard and Onco-mouse (1991) Eur. Pat. Off. Rep. 525.

[81]U.S. Patent No. 4,736,866,1089 Off.Gaz.Pat.Off.703(April 12, 1988).

non-human mammal' as extending to an animal variety.[82] Plant or animal varieties
are even be subsumed themselves under a patent if the variety is produced by non-
biological methods and the application requests a product by process patent.
Again, the US pioneered in this respect.[83] The EC Directive on patent protection
for biotechnological inventions appears to follow this line.[84] Moreover, the Art. 53
(b) EPC exemption for biological methods of producing new plant or animal varieties is more and more narrowly interpreted in order to extend the patentable space
for genetic engineering. According to the said EC Directive, a method is to be
regarded as non-biological (and hence patentable) already if, in step-by-step methods, only one step is based on genetic engineering.[85] Thus, overall, the drive enabling patenting in biotechnology has even invaded those realms which were
exempted to pay tribute to the givenness of life (Rimmer 2008: 26).

## 7.3.2  Patent—A Stimulus of What?

Apart from the discourse about patenting life forms, doubts about the functions of
patent law in general can also be raised. Critiques have argued that the broad patenting of DNA sequences and proteins, of techniques of analysis, and of bioinfor-
matic tools managing and combining the data-masses, hinders free research rather
than stimulating it. This may be the case especially in synthetic biology where the
number of bioparts that are used and therefore require patent licence and royalty
payment may rise to a level that certain projects become difficult to manage.[86]
That has prompted some synthetic biology promotors to advocate for open source
policies (cf. Holman 2015: 427). The hinderance of R&D they fear has been
described as a tragedy of the anti-commons. It holds that, other than in the tragedy
of the commons where common use consumes the common good, in the tragedy
of the anti-commons the privatisation of the common good hinders the further
development of the same (Rimmer 2008: 9).

   It is however difficult to prove if that is true. After all, the USA has since ever
excelled in their share in the worldwide patent applications. Of all applications in

---

[82]EPO Technical Appeal Chamber, decision of 3 Oct 1990 in GRUR (1990), 978. The argument
is very positivistic because by Art. 53 (b) EPC it was meant to exclude patenting of living beings
(be they varieties or genera or whatever). On the other hand, at the times of negotiations of the
EPC one could not foresee how much biotechnology would become able to cut living forms into
pieces and recombine them.

[83]Ex parte Hibberd, 227 USPQ 473 (PTO Bd. Pat. App. & Int., 1985).

[84]Art. 4 (3) Directive 98/44/EC.

[85]Cf. Art. 2 (2) Directive 98/44/EC.

[86]Consider, for instance, the following statement of the BioBricks Foundation. Referring to the
MIT Registry of Standard Biological Parts it is said: "There are over 10,000 parts in this reposi-
tory, and it keeps growing. To patent each of these parts would already cost tens of millions of
dollars: if you gave a would-be engineer of biology that much money he or she would probably
use it to make better parts." See https://biobricks.org/bpa/faq/#6 (accessed 27.07.2015).

biotechnology in 2006 globally 41.5 % came from the US (van Beuzekom and Arundel 2009: 71). This is hardly an indication that the research in the USA was hindered by patents. But, from the inverse perspective, this is also not a proof that the US-based research has been more beneficial for mankind. I would rather argue that patenting practices reflect differences of industrial strategies between countries. For instance, the share of the EU in patent applications in biotechnology in 2006 was only 27.4 %, although the number of biotechnology firms was larger in the EU than in the US.[87] This may reflect an attitude of researchers in the EU to consider their results to be due to the public domain. The US R&D in biotechnology may nevertheless have been more dynamic due to the patentability of life forms,[88] at least until the Supreme Court took its restrictive turn in *Mayo* and *Myriad*.[89] But the crucial point is, I believe, not the dynamics as such but the direction it takes. Patenting directs R&D towards commercializability rather than towards the generation of basic knowledge, knowledge for the public good, and knowledge about risks. It appears that this kind of knowledge flourishes better if generated in the public domain. It may grow at a slower pace but its pluralistic nature helps to avoid detrimental trajectories. For instance, if universities nowadays push for patenting this is a serious assault on the public domain of which they claim to be an important part. The public research domain has since long been well guided by the stimulus of scientific esteem for researchers. It will lose its independence from short term economic goals if esteem is replaced by financial gain as the major incentive.

In conclusion, it appears that patenting in biotechnology has rather become a means of rushing to seize and appropriate as much valuable R&D results as possible in order to exclude others.[90] The initial idea of stimulating R&D for the public good seems to have been replaced by a scheme where a few fast runners make the rest of the R&D world pay royalties. The big R&D players get hold of an ever-increasing part of life. In addition, enormous transaction costs have emerged on all sides—patent holders, government, competitors, opposing NGOs—which are hardly justified by the benefits for society. Therefore, when exploring the details of patent preconditions and content a critical stance is appropriate, attempting to apply patent law with a view to avoid distributive injustice and premature commercial aspiration.[91] This implies, in particular, that the preconditions of patenting should be interpreted restrictively and with a view to shifting patenting from the early to later phases of the R&D process.

---

[87]Namely 3301 firms in the US and 3377 firms in the EU, not counting the Member States from which data were not available, see van Beuzekom and Arundel (2009) p. 15.

[88]Mueller (2010) p. 230 who points to the fact that the Chakrabarty decision of the US Supreme Court has been a strong stimulus for the biotechnology industry.

[89]See above Footnote 79.

[90]See Holman (2015) p. 389 et seq for a description of the patent strategies of Genentech, Amgen and Monsanto.

[91]See for a similar approach Rimmer (2008).

### 7.3.3  Applying Patent Law to Synthetic Biology

At first sight the concerns about patenting living nature have less grounding in the context of synthetic biology because of its increasing degree of artificiality of processes and products. However, the opposite may also be true: First, until today life has not been artificially constructed. Any organism originating from synthetic biology has as yet been one that took its life from nature, not from man. Secondly, the ever increasing artificiality suggests that the technological development should not be accelerated but slowed down in order to allow for time for sound evaluation of development lines. If patents were refused on phenomena of nature and life forms, or if they were granted not prematurely but at a late stage in the R&D process, and more restrictively, more room would be available for a pluralist generation of ideas and evaluative discourses. This would prevent path dependencies emerging, which make it difficult to suppress the bringing on the market of unneeded or detrimental products at a later stage.

In that line open source systems have been set up in synthetic biology. A major example is the Biobrick Foundation. It runs a data bank to which anyone can contribute standardised biobricks and which can be used by anyone. The contributor is asked not to assert any IPR against a user while the user must agree to attribute publications and products to the data bank and—if she so requests—the contributor.[92]

In summary, supporting the public domain against patenting may be justified on two levels: On the level of "objective" nature, because the "technology" of nature belongs to humanity as a commons,[93] and on the level of "subjective" human ingenuity because the "technology" of humankind is a collective undertaking which best flourishes in the open exchange of ideas[94].

A more patent friendly way is to suggest a restrictive approach to understanding and applying the patenting preconditions and the protective scope of a patent.[95]

Four preconditions must be met for a patent: The object of patenting must be an invention in a field of technology, the invention must be novel, it must involve an inventive step, and it must be susceptible to industrial application.[96]

---

[92]The BioBrick Contributor Agreement and the BioBrick User Agreement are available at https://biobricks.org/bpa/ and http://dspace.mit.edu/handle/1721.1/50999 (accessed 27.07.2015).

[93]Consider the statement of Charles Darwin at the beginning of this chapter.

[94]For an analysis see Torrance (2010) pp. 659 et seq.

[95]On the important question what agent would be more prepared to adopt a new policy—the patent administration and its networks, the legislator or the judiciary—see Schneider (2014).

[96]See Article 52 (1) EPC: "(1) European patents shall be granted for any inventions, in all fields of technology, provided that they are new, involve an inventive step and are susceptible of industrial application".

Concerning the notion of invention in the field of biotechnology gene sequences, proteins, cells and microorganisms are all patentable objects, whether natural or genetically engineered. However, it is important to stress that the applicant must prove a function of the object.[97] It is not sufficient to just describe it, even if in artificially purified form (Mueller 2006: 227). Otherwise genomes, cells and microorganisms would soon all become private property. But even if it is ensured that genes, cells and microorganisms are only patented if the function has been identified, this would still be an obstacle for pluralist research. Therefore, reconsidering the fact that the genetic resource and its function is not an invention but a discovery, no natural gene, cell or microorganism should be patentable at all. Patents should only be allowed on modified constructs and the method of their construction.

Concerning the novelty requirement, the notion of state of the art is crucial as reference.[98] The state of the art comprises making the invention publicly available by means of oral or written description, by use or in any other way. In order to be relevant to the assessment of novelty, the prior publication must be enabling. This means that the mere description of a gene does not hinder the patentability of a later description of the function encoded in the gene (provided a discovery is accepted as patentable at all).

Data base entries of nucleotide sequences have been considered as representing state of the art, if the data bases are of a kind:

(a)  which are known to the skilled person as an adequate source for obtaining the required information,
(b)  from which this information may be retrieved without undue burden, and
(c)  which provide it [the information] in a straightforward and unambiguous manner without any need for supplementary searches.[99]

The case leading to this general statement was a request concerning chimeric genes useful for mediating herbicide tolerance in plants. The relevant coding sequence had been deposited in a publicly available database which was considered to represent common knowledge. Thus, as the public domain of R&D results destroys novelty, the willingness of the R&D community to make results publicly available is crucial for the realm of patentability.

---

[97]This is stated in consideration no. 23 of Directive 98/44/EC:" Whereas a mere DNA sequence without indication of a function does not contain any technical information and is therefore not a patentable invention". The function must also be described as being industrially applicable, see consideration no. 24 of Directive 98/44/EC which reads: "Whereas, in order to comply with the industrial application criterion it is necessary in cases where a sequence or partial sequence of a gene is used to produce a protein or part of a protein, to specify which protein or part of a protein is produced or what function it performs." However, these restrictions are unfortunately not expressed in the wording of the patent preconditions.

[98]Art. 54 EPC. On case law see Jaenichen et al. (2012), Chaps. 15 and 16.

[99]EPO T 890/02–3.3.8, "Chimeric gene/Bayer", headnote; Jaenichen et al. (2012), p. 610.

Concerning the precondition of an inventive step, it is current practice to identify the closest prior art and compare the invention with this, considering whether
the inventive step was not obvious to a person skilled in the art.[100] A restrictive
interpretation would particularly target the reference person searching for her in
the global public domain and assuming her to have some skill. Likewise, what is
"obvious" might be interpreted very broadly.[101]

Concerning the precondition that the invention must be susceptible to industrial
application, critique has been raised against the current practice that information
is already patented if its application is only aimed at further R&D. For instance,
a patented gene function which needs further R&D in order to be developed into
a product is already considered to be industrially applicable. A researcher who
wishes to explore it further and develop a product from it must pay royalties. It is
submitted that "industrial applicability" should be interpreted more restrictively,
such that only products ready to be brought on the market are patentable.

From a more radical perspective it can be alleged that industrial applicability
should be understood to mean use value, not exchange value. This would mean
that it does not suffice if a product can earn money, but that some substantial benefit for society is to be expected (Winter 1992: 185).

Concerning the scope of the patent protection it has been debated whether the
protection is absolute in the sense that any further use of a patented DNA sequence
is subject to the consent of the patent-holder, or whether the protection should be
restricted to the patented function and its industrial applicability thus allowing for
free use for other functions. It is suggested that the latter approach should be taken
in order to limit the power of the patent holder to steer the further technological
development.[102]

### 7.3.4 Copy Right for Synthetic Biology Programs?

It has been suggested that a copy right might be more appropriate than a patent
right as intellectual property in the area of synthetic biology (Holman 2015:
458–462). Indeed, the design of new organisms has more and more become an
exercise in silico rather than experimentation. It very much resembles a computer

---

[100]Art. 65 EPC; Jaenichen et al. (2012), Chaps. 17 and 18.

[101]The EPO appears to take a more patent friendly stance which does not take note of the concern expressed here. See the summarizing formulations in EPO (2013), p. 184: "In accordance
with the case law of the boards of appeal, a course of action could be considered obvious within
the meaning of Art. 56 EPC if the skilled person would have carried it out in expectation of some
improvement or advantage (T 2/83, OJ 1984, 265). In other words, obviousness was not only at
hand when the results were clearly predictable but also when there was a reasonable expectation
of success (T 149/93)".

[102]For an elaborate discussion of the absolute or function-orientated protective scope see
Kunczik (2007) pp. 156–193.

program, which has been included in the copyright regime.[103] With the preconditions of copyright (such as originality, authorship, creation, work, etc.[104]) it would
be clear that no programme found in nature is protectable. However, the analogy
stops short in some important respects: synthetic biology computer programmes
do not aim at being operated within the in silico world but shall be translated into
the organismic world. The protection must therefore extend to that real world.
Copyright however confers the exclusive right of publication of the programme
but not the making use of its content. Moreover, the copyright originates without
administrative oversight and thus lacks legal certainty, while a patent right must be
granted by administrative decision. Besides, the time span of protection[105] is much
too long. Therefore, if an IPR shall be at all, a patent right—a reformed version, to
be sure—appears to be the better solution.

## 7.4  Redistributing Benefits from Synthetic Biology

Targeted breeding, genetic engineering and synthetic biology all work on the
same resource: the genetic program of organisms. Even the most advanced techniques of synthetic biology, such as the replacement of the complete genome of a
microorganism, is still based on the life processes of the original organism. In the
Venter experiment, notably, the synthetic genome depended on the natural apparatus of the recipient cell, its "machinery of transcription, translation, and replication, its ribosomes, metabolic pathways, its energy supplies, and so on." (Church
and Regis 2012: 50). This means the vital power was not artificially created but
used and redirected. And even the imported genome—although synthesized—was
mainly a replication of the real genome of the donor organism, with some newly
designed strains of minor importance.

The fact that the genetic programs are still the physical and intellectual origin
of synthetic biology raises the question if there is ownership in those programs and
if that is the case, what bearing ownership would have.

In fact, after some debates about the best solution the Convention on Biological
Diversity (CBD) has determined that states have sovereign rights over their genetic
resources and hence the genetic programmes.[106] In 2014 the Nagoya Protocol (NP)
entered into force, which specifies the basic CBD rules further. The object of the
sovereign rights is termed "genetic resources" which are defined as genetic material

---

[103]In the EU see Directive 2009/24/EC on the legal protection of computer programs.

[104]The criteria vary according to national legislation. They have hardly been internationally
harmonized. In Germany computer programs are protected if involving a "genuine intellectual
creation" (eigene geistige Schöpfung), see Sect. 69a Urheberrechtsgesetz (Copy Right Act).

[105]It varies according to national legislation. In Germany it extends to 70 years after the death of
the author.

[106]Art. 15 CBD.

of plant, animal, microbial or other origin containing functional units of heredity and having actual or potential value.[107] Geographically the sovereign rights of states extend to "their"[108] genetic resources, which means to those found within a state's territory, including the territorial sea, the exclusive economic zone and the continental shelf. Content-wise, the sovereign rights comprise the right to regulate access to the genetic resources and the right to claim the sharing of benefits drawn from them.[109] The state allowing access can use its right to determine, by administrative act and/or a contract with the accessor, the kinds of genetic resources allowed to be sampled or otherwise acquired, the R&D allowed to be performed, the use of such results for commercial purposes or the public domain, etc. As for benefit sharing, monetary benefits (for example from the marketing of products) and non-monetary benefits (for example the sharing of research results and the participation of provider state personnel in R&D projects) are possible.[110] The state can, when negotiating the access agreement with the researcher (called mutually agreed terms in the CBD and NP language), ask for either or both of these benefits.

Prior consent of the provider state is not necessary if this state has waived its right to establish an access regime. Many states have opted this way, especially states of the industrialised world. They allow any researcher to take a sample and use it freely for any R&D.

Where an access regime has been established two questions concerning benefits arise. The first is whether the influence of the genetic resource can wither away in the process of valorization until a final product. This depends on the interpretation of the wording in Art. 15.5. CBD, "benefits arising from the commercial and other utilization of genetic resources", and Art. 5.1 Nagoya Protocol, "benefits arising from the utilization of genetic resources as well as subsequent applications and commercialization." Do benefits still "arise from" "commercialization" in the following exemplary case? A gene is extracted from an organism X accessed in state A. Its genome is sequenced. The DNA data are put into a public database. The data are taken from the database. By computer screening or tests, the functions encoded in the gene are identified. The result is again put into a public database. The functional data are used to plan the modification of an organism Y. The gene is inserted into organism Y. The modified Y is multiplied and sold on the market. It generates profitable revenue. Can one say that in the course of valorization the contribution of the original genetic resources is lost? Should the case be treated in analogy to the exhaustion of a patent right which loses its power if an invention is achieved that is novel compared with the original invention? These questions are still unsolved and rarely even discussed. My own reaction is that as long as

---

[107]Definition compiled from Art. 2 paragraphs 9 and 10 CBD.

[108]See Art. 15.1 CBD.

[109]Art. 15 CBD and Arts. 5 and 6 Nagoya Protocol.

[110]Art. 15.5 CBD speaks of "sharing in a fair and equitable way the results of research and development and the benefits arising from the commercial and other utilization of genetic resources".

the original genetic program is still present in the final product its coverage is not exhausted. The case would be different if the genetic program were only used as a means of comparing genes and the result—the identification of a function—related to the gene from organism Y rather than to that from organism X.

The second question is whether not only natural but also artificially modified genetic resources should be subject to the sovereign rights of states. For instance, if a German researcher applying synthetic biology methods constructs a new microorganism in a Brazilian laboratory, has Brazil sovereign rights to regulate the access and utilization of this organism, allowing the state, for instance, to ask for prior consent? In more general words, is any new genetic resource the "property" of the state? Once more, the question is hard to answer and has hardly even been discussed. My suggestion is this: The state has indeed sovereign rights over those new creations. This follows from the definition of genetic resources which covers genetic material not only of plant, animal and microbial origin, but also of "other origin".[111] It would however be wise for any state not to make use of its regulatory powers in such cases. If the state says nothing about property in artificial genetic programs, these simply are ownerless. The only kinds of property rights affecting them is ownership of the natural genetic program contained in the artificial one, as well as any patent rights or other intellectual property rights that may be obtained.

In conclusion, it appears that the upcoming legal regime of access to genetic resources and benefit sharing may have consequences for synthetic biology, but that this question has hardly touched the relevant R&D community, nor governments responsible for the sector.

## 7.5 Summary

Law in modern societies approaches new scientific and technological developments in three characteristic ways: It provides forms which enable innovation and at the same time sets limits in order to prevent unwished side-effects; it provides incentives that foster R&D development and steer it into certain directions; and it sees to that benefits from R&D results are fairly shared. Synthetic biology is also exposed to this legal interference.

(1) Law enabling and regulating synthetic biology

The current dynamic evolution of synthetic biology research is enabled by the constitutional freedoms of scientific research as well as of free enterprise and property. These rights allow for regulatory restrictions in the public interest. Such regulation must be introduced if constitutional rights (such as of health) of third persons or constitutional obligations of environmental protection so require.

---

[111]Art. 2 (9) CBD.

In view of the fundamental change in socio-economic life that must be expected from synthetic biology, it would be politically wise to develop a legal basis for it. This legal basis should take a fresh look at the entire biotechnology and establish a framework not only for synthetic biology but also for genetic engineering and high tech breeding of plants and animals. Constitutional law can be interpreted to enable and even require such legal framework, because biotechnology affects fundamental rights and obligations, and because it encroaches on the separation of powers which assumes that "essential" innovations should be deliberated by parliaments rather than be left to the executives.

Alternatively, the regulation of synthetic biology could also be based on the current legislation on genetically modified organisms. However, although some strands of synthetic biology falls under the definition of genetic modification others are not. They should be included in the scope of application of GMO laws.

More precisely, the following synthetic biology techniques are covered by the legal definition of GMOs:

- Techniques of the first strand of synthetic biology, here called radical genetic engineering, unless they create a completely new organism; for instance, the partial replacement of the content of a cell with synthesized material of natural and artificial design would still be regarded as the production of a GMO because an original cell (and its capacity to live) was used.
- Xenobiology insofar as it incorporates genetic xeno-material into an existing organism.

In contrast, the following synthetic biology activities are not captured by the GMO regime:

- Within the first strand:
  - the full synthesis of an organism and the complete replacement of the genetic material of an organism, be it of known or new design, including xenomolecules
  - the synthesis and placing on the market of bioparts

- The second strand, i.e. the designing and synthesis of a protocell or minimal cell (be it constructed top-down or bottom-up)
- Within the third strand: xenobiochemistry, insofar as it alters chemical derivatives (amino acids, proteins) of an organism thus producing a chemically modified organism (CMO).

Even if the non-captured kinds of synthetic biology were subsumed to the GMO regime the current methodology of risk assessment, which was developed for the genetic modification of organisms, must be examined as to its fit to synthetic biology. Considering the increasing depth of intervention and artificiality of constructions the familiarity principle used in GMO risk assessment must be replaced by a methodology which better accounts for new designs.

Considering the impossibility of proving a zero risk to health or the environment more criteria of risk evaluation should be applied. If a risk assessment concludes that the risk is low, the project should nevertheless be asked to prove that it provides a benefit to society. Moreover, cultural concerns of populations about the benefits, risks and ethics of synthetic biology should be taken seriously in the regulatory discourse and not regarded as irrational and therefore negligible. They should, if they can be substantiated, be accepted as legitimate ground for regulatory restrictions.

For the time being until the regulatory regime will be established a moratorium should be applied on any release into the environment of organisms resulting from synthetic biology.

(2)  Law fostering synthetic biology

Intellectual property discourse and legislation have since a long time disapproved of the argument that natural phenomena should not be appropriated by private persons. It was acknowledged that discoveries and technical advancements of natural phenomena shall be patentable. This privatisation of nature is now being extended to the results and methods of synthetic biology. Although this technology is still very much dependent on natural processes, material and information, the privatization seems to have more justification because of the increasing proportion of technical input. However, this proportion is still poor in relation to the "higher workmanship of nature" (Darwin). Life and the natural processes of its evolution are a common good which should not be appropriated by anyone. Moreover, the privatization of life forms has more and more hindered pluralistic R&D. There is a risk that the direction of R&D will gradually be determined by commercial objectives, and that they will hinder basic research, research on risks and research on benefits for society as a whole. For this reason open source systems have emerged. They deserve to be supported. In addition, the patenting preconditions should be restrictively interpreted so that patents are not provided for information accruing in the early stages of the R&D processes but only at the end when products are ready to be brought on the market.

(3)  Law ensuring benefit sharing

Synthetic biology has hardly taken notice of the fact that it will be affected by an emerging regime introduced by the CBD and specified by the Nagoya Protocol which acknowledges sovereign rights of states to regulate access to their genetic resources and ask for a share in the benefits drawn from R&D on them. This appropriation of genetic resources by resource states is a response of developing states to IPR strategies of industrialized states. This is understandable although it contradicts the idea of common goods. The regime suggests a bilateral exchange, "access to genetic resources for a share in commercial revenue from R&D". This may work if the R&D process is short. It is problematic, if the valorization chain is long and complex because it is then hardly possible to trace benefits back to a genetic resource that was sampled in a specific provider state. Therefore, common pools should be tried as solutions. They would consist of cooperation between users and providers of genetic resources in R&D activities and subsequent commercialization. This option is preferable because the provider state, by participating in the whole process, stays informed about its incurring revenues and—even more importantly—is able to build up its own R&D capacity. Concerning synthetic biology based new organisms the question arises if the state in which it is located can claim sovereign rights over them. This is arguably the case. But it is submitted that states should desist from introducing an access and benefit sharing regime in this respect.

# References

Acatech, Leopoldina and Deutsche Forschungsgemeinschaft (2009), Synthetische Biologie. Stellungnahme, Wiley-VCH Verlag, Weinheim

Baldwin G, Bayer T, Dickinson R, Ellis T, Freemont PS, Kitney RI, Polizzi K, Stan G-B (2012) Synthetic biology. A primer. Imperial College Press, London

Bölker M (2015), Complexity in synthetic biology: unnecessary or essential? In: Giese B, Pade C, Wiener H, von Gleich A (eds) Synthetic biology – character and impact. Springer, Berlin, pp 59–69

Breckling B (2004) Naturwissenschaftliche Grundlagen der Gentechnik als Ausgangspunkt zur Risikoabschätzung gentechnisch veränderter Organismen, in: Umweltbundesamt (ed.) Fortschreibung des Konzepts zur Bewertung von Risiken bei Freisetzungen und dem Inverkehrbringen von GVO, Berichte 3/04, E. Schmidt Verlag, Berlin

Breckling B, Schmidt G (2015), Synthetic biology and genetic engineering: parallels in risk assessment. In: Giese B, Pade C, Wiener H, von Gleich A (eds) Synthetic biology – character and impact. Springer, Berlin, pp. 197–211

Budisa N (2012) Chemisch-synthetische Biologie. In: Köchy K, Hümpel A (eds) synthetische Biologie. Berlin-Brandenburgische Akademie der Wissenschaften, Berlin, pp 85–115

Bühler L (2006) Besonderheiten von biotechnologischen und computerimplementierten Erfindungen. In: von Büren R, David L (eds) Schweizerisches Immaterialgüter und – Wettbewerbsrecht. Helbing & Lichtenhahn, Basel, pp. 509–513

Bundesregierung (2011) Antwort der Bundesregierung auf eine Kleine Anfrage "Stand und Perspektiven der Synthetischen Biologie" of 18 March 2011, BT-Drs. 17/4898

Church G, Regis E (2012) Regenesis. How synthetic biology will reinvent nature and ourselves. Basic Books, New York

Correa CM (2000) Intellectual property rights, the WTO and developing countries – The TRIPS agreement and policyoptions, London, New York

Danchin A (2015) The cellular chassis as the basis for new functionalities – shortcomings and requirements. In: Giese B, Pade C, Wiener H, von Gleich A (eds) Synthetic biology – Character and impact. Springer Berlin, pp. 155–172

Darwin C (1859) Origin of species. Caldwell, New York

Danner K (1987) Bedürfnisse der Anmelder biotechnologischer Erfindungen. GRUR 1987:315

Dolezel M, Miklau M, Eckerstorfer M, Hilbeck A, Heissenberger A, Gaugitsch H (2009) Standardising the environmental risk assessment of genetically modified plants in the EU. Bundesamt für Naturschutz, Bonn, BfN-Skripten 259/ 2009

EPO (European Patent Office) (2013) Case law of the board of appeals of the EPO. EPO. http://documents.epo.org/projects/babylon/eponet.nsf/0/4CCF968D57494023C1257C1C004F992C/$File/case_law_of_the_boards_of_appeal_2013_en.pd

French CE et al. (2015) Beyond genetic engineering: technical capabilities in the application fields of biocatalysis and biosensors. In: Giese B, Pade C, Wiener H, von Gleich A (eds.) Synthetic biology – character and impact. Springer, Berlin, pp. 113–137

Friends of the Earth, CTA and ECT (2012) The principles for the oversight of synthetic biology http://www.foe.org/news/blog/2012-03-global-coalition-calls-oversight-synthetic-biology Accessed 06 Jan 2014

Giese B, von Gleich A (2015) Hazards risks and low hazard development paths of synthetic biology. In: B Giese, Pade C, Wiener H, von Gleich A (eds) Synthetic biology – character and
impact. Springer, Berlin, pp 173–195

Gibson DG (2010) Creation of a bacterial cell controlled by a chemically synthesized genome.
Science 329:52. doi:10.1126/science.1190719

Goldstein JA (1987) Der Schutz biotechnologischer Erfindungen in den Vereinigten Staaten,
Gewerblicher Rechtsschutz und Urheberrecht, pp 310–317

Holman CM (2015) Developments in synthetic biology are altering the IP imperatives of biotechnology. Vand J Ent Tech L 17(2): 385–462.

Jaenichen JA, McDonnell LA, Haley, JF, Hosoda Y, Meier J (2012) From clones to claim. The
European patent office's case law on the patentability of biotechnology inventions in comparison to the United States and Japanese Practice, C. Heymanns, Köln

Kamau E, Winter CG (2015) Research and development on genetic resources. Public domain
approaches in implementing the Nagoya protocol. Routledge, Oxon

Kunczik N (2007) Geistiges Eigentum an genetischen Informationen. Nomos, Baden-Baden

Leduc S (1912). La Biologie Synthétique. A. Poinat, Paris. http://www.peiresc.org/bstitre.htm

Lynch D, Vogel D (2001) The regulation of GMOs in Europe and the United States: a case-study
of contemporary European regulatory politics, Council of Foreign Relations, Washington.
http://www.cfr.org/agricultural-policy/regulation-gmos-europe-united-states-case-study-con-
temporary-european-regulatory-politics/p8688. Accessed 14 Feb 2014

Mandel GN, Marchant GE (2014/2015) The living regulatory challenges of synthetic biology.
Iowa L Rev 100:155–200

Maurer SM (2010/2011) End of the beginning of the end? Synthetic biology's stalled security
agenda and the prospects for restarting It. Val. U. L. Rev. 45:1387–1446

Mills O (2010) Biotechnological inventions. Moral restraints and patent law. Ashgate, Burlington

Mueller JM (2006) An introduction to patent law. Aspen, New York

Neumeier H (1990) Sortenschutz und/oder Patentschutz für Pflanzenzüchtungen

Paradise JD, Fitzpatrick E (2012–2013) Synthetic biology: does re-wrtiting nature require re-
writing regulation? Penn St. L. Rev. 117:53–88

Parisi C (2012) new plant breeding techniques. State of the art. potential and challenges. Servicio
de Publicaciones de la Universidad de Cordoba. http://helvia.uco.es/xmlui/bitstream/handle/
10396/9492/2013000000688.pdf?sequence=1.Accessed 6 Jan 2014

Raaijmakers M (2009) What are the rules on cell fusion techniques in the EU public laws and
private standards for organic farming. In: ECO-PB Workshop (2009), Strategies for a future
without cell fusion techniques in varieties applied in organic farming. http://www.ecopb.org/
fileadmin/ecopb/documents/Proceedings_Paris_090427.pdf. Accessed 6 Jan 2014

Rimmer M (2008) Intellectual property and biotechnology. Edward Elgar, Cheltenham

Schmidt M (2010) Xenobiology. A new form of life as the ultimate biosafety tool. Bioessays
32(4):322–331

Schneider I (2014) Exclusions and exceptions to patent eligibility revisited: examining the political functions of the "discovery" and "order public" clauses in the European patent convention and the Arenas of negotiation. In: de Miguel Beriain I, Cassabona CMR (eds) Synbio
and human health. A challenge to the current IP framework?. Springer, Berlin, pp 143–173

Schummer J (2011) Das Gotteshandwerk. Die künstliche Herstellung von Leben im Labor.
Suhrkamp, Berlin

Spök A (2010) Assessing socio-economic impacts of GMOs, Wien (Bundesministerium für
Gesundheit). http://www.bmg.gv.at/cms/site/attachments/2/7/6/CH0808/CMS1287125505520/
assessing_socio-economic_impacts_of_gmos,_band_2_2010.pdf

Torrance AW (2010) Synthesizing law for synthetic biology. Minn J Law, Sci Technol
11(2):629–665

US Patent and Trade Mark Office (2014) Procedure for subject matter eligibility analysis of
claims reciting or involving laws of nature/natural principles, natural phenomena, and/or
natural products, Memorandum March 4, 2014

van Beuzekom B, Arundel A (2009) OECD biotechnology statistics 2009, OECD. http://www.oecd.org/sti/sci-tech/42833898.pdf. Accessed 19 Jan 2014

van Doren D, Koenigstein S, Reiss T (2013) The development of synthetic biology: a patent analysis. Syst Synth Biol. DOI 10.1007/s11693-013-9121-7

von Danwitz T (2008) Europäisches Verwaltungsrecht, Springer, Berlin

von Kries C, Winter G (2011) The structuring of GMO release and evaluation in EU law. Biotechol J 7(4):569–581. doi:10.1002/biot.201100321

Winter G, Mahro G, Ginzky H (1993) Grundprobleme des Gentechnikrechts, Düsseldorf (Werner Verlag) 1997

Winter G (1992) Patent law policy in biotechnology. J Environ Law 4(3):167–187

# Legal Acts cited

Commission Decision 2000/608/EC of 27 September 2000 concerning the guidance notes for risk assessment outlined in Annex III of Directive 90/219/EEC on the contained use of genetically modified micro-organisms, Annex, OJ 2000 L 258, p 43

Commission Decision 2002/623/EC of 24 July 2002 establishing guidance notes supplementing Annex II to Directive 2001/18/EC of the European Parliament and of the Council on the deliberate release into the environment of genetically modified organisms and repealing Council Directive 90/220/EEC, OJ 2002 L 200, p 22

EFSA Panel on Genetically Modified Organisms (GMO), Guidance on the environmental risk assessment of genetically modified plants. EFSA J. 2010, 8, 1879. Accessible via www.efsa.europa.eu/efsajournal.htm

Directive 98/44/EC of the European Parliament and of the Council of 6 July 1998 on the legal protection of biotechnological inventions, OJ 1998 L 213, p 13

Directive 2001/18/EC of the European Parliament and the Council of 12 March 2001 on the deliberate release into the environment of GMOs […], OJ 2001 L 106, p 1

Regulation (EC) No 1829/2003 of the European Parliament and of the Council of 22 September 2003 on Genetically Modified Food and Feed (Text with EEA relevance), OJ 2003 L 268, p. 1

Directive 2004/35/EC of the European Parliament and of the Council of 21 April 2004 on environmental liability with regard to the prevention and remedying of environmental damage, OJ 2004 L 143, p 56

Directive 2009/24/EC of the European Parliament and of the Council of 23 April 2009 on the legal protection of computer programs, OJ 2009 L 111, p 16

Directive 2009/41/EC of the European Parliament and of the Council of 6 May 2009 on the contained use of genetically modified micro-organisms (Recast) (Text with EEA relevance), OJ 2009 L 125, p 75

Printed in the United States
By Bookmasters